Microeconomic Studies

Edited by W. Güth, J. McMillan and H.-W. Sinn

Michael Funke (Ed.)

Factors in Business Investment

Papers of a Conference Held at the Science Centre, Berlin, Research Area "Labour Market and Employment", September 1987

Springer-Verlag
Berlin Heidelberg New York
London Paris Tokyo

Dr. Michael Funke
Science Centre Berlin
Research Area, Labour Market and Employment
Reichpietschufer 50, D–1000 Berlin 30, West-Germany

ISBN-13: 978-3-642-48750-7 e-ISBN-13: 978-3-642-48748-4
DOI: 10.1007/978-3-642-48748-4

Softcover reprint of the hardcover 1st edition 1989

2142/3140-543210 - Printed on acid-free paper

Preface

The papers in this volume were originally presented and discussed at the conference "Factors in Business Investment" held at the Science Centre Berlin on September 21-23, 1987. All papers are published subject to some revision made in the light of discussants' formal comments and the lively debate during the conference sessions.

The main aims of the international conference were to discuss current modelling of investment behaviour, to extend our understanding of certain key elements in the problem and to discuss the appropriate policy responses. A central tenet of this initiative was the conviction that the interchange of ideas across national boundaries, particularly among economists attempting to analyse current problems with modern theoretical and quantitative techniques, could both contribute to a better understanding of the problems and provide an effective springboard for further research. A fuller description of the background to the conference and what it achieved is provided in the introduction.

My thanks go to all the participants who shared with me the three days of intensive discussions. They created an atmosphere which I recall as being extraordinarily stimulating for an academic exchange of views on the frontiers of theoretical and empirical research. I am also extremely grateful for the financial support of the Science Centre Berlin and the Fritz Thyssen Foundation. Furthermore, I would very much like to thank the editors of this series Werner Güth, John McMillan and Hans-Werner Sinn.

September 1988 Michael Funke

Table of Contents

Introduction: A Summary of the Issues

Michael Funke

Science Centre Berlin

The level, the composition, the growth and the efficiency of investment is considered to be the main determinant of future economic growth and a key to the revival of output and employment in the OECD countries. Indeed, it has often been stressed that sustained economic recovery depends on renewed growth of fixed investment, and various recent economic analyses have been concerned with the conditions for such a renewal. In view of this importance it is obvious that there is a marked increase in serious economic research about factors affecting investment decisions. Inevitably, approaches to economic analysis and individual points of emphasis have displayed a wide heterogeneity of interest.

Existing empirical work on investment behaviour is usually based on accelerator theories, the Jorgenson approach, or Tobin's Q-theoretic framework. Most investment equations presented in the papers in this volume are straightforward modifications or blends of these three theories of investment demand. Because all three types of models have been discussed in the literature in great detail only the main characteristics are covered here.

The accelerator model proposes that firms' demand for investment goods depends on changes in the demand for business products. The firm's stock of capital varies directly with its level of production, so changing demand will induce the firm to alter its productive capacity. The reaction of the capital stock to changes in demand may be lagged, due to lags in delivery of investment goods. In accelerator models an increase in the demand for investment goods must pass through stages of planning and ordering before intentions become expenditures. The gradual response to changes in demand which is represented by the sequence of lagged output terms also serves for another purpose. Because capital goods are long-lived, investment must depend upon expectations of future demand as well as current demand. If future output expectations are extrapolated from past sales patterns, the lagged output terms in the investment equation also represent output projections which are derived from a "backward-looking" expectations scheme.

An important omission of the accelerator theories of investment has been signalled by the neoclassical model of investment behaviour pioneered by Jorgenson: Investment is assumed to be independent of the development of factor prices. Jorgenson introduced these explicitly into an investment model which is based on the assumption of a profit maximizing firm. The so-called neoclassical model assumes perfect competition on all markets and the absence of adjustment costs. Under the assumption of perfect foresight and a feasible set of production possibilities given by a well-behaved production function firms will invest until the marginal cost of capital is equal to the net present value of the future returns on marginal investment. The cost of capital in this model consists of five elements: First the interest rate, second the depreciation rate, third the taxation of capital, fourth the price of the capital good, and, finally, the change of the price of capital goods. In order to dynamise the neoclassical model, lags are normally introduced. While the model rigorously derives the firm's desired capital stock from a profit maximization problem, it is ad hoc in the specification of the dynamics of the investment process. These lags may be justified on the same grounds as in the accelerator models: delivery lags or lags in expectations. Given these characteristics the accelerator model can be considered as a restricted form of the neoclassical model, with zero elasticity of substitution between capital and labour. This assumption implies that relative factor prices have no influence on investment, i.e. the Jorgenson model collapses into an accelerator model.

Tobin's Q-theory of investment which has been used as frequently as accelerator or neoclassical models in empirical research considers the ratio of the market value of real capital assets to its reproduction cost as the primary determinant of investment. The market value of capital assets may be seen as a market estimate of expected profits, and so may be interpreted as the demand price of capital. The replacement cost of reproducible real capital assets is an indication of the supply price of capital. So, Tobin's Q which proxies the incentive to acquire new capital is the ratio of the demand and supply price of capital. If the valuation ratio is above its equilibrium value (normally taken as unity), then a profit maximizing firm has an incentive to invest, as the discounted profit stream generated by the new capital will be greater than the cost of acquiring it. This implies that the stock market should be a good predictor of the rate of private investment. The assumption that investment depends on Q is consistent with the adjustment cost theory. If the demand price is the sum of the price of capital goods and adjustment costs, and if adjustment costs are an increasing function of investment, then investment is determined solely by Q. An appealing feature of the Q theory is its explicit consideration of "forward-looking" expectations. These enter the investment equation directly through the market value of the firm taken from stock market data, incorporating, in principle, the expectation of all future variables relevant to the investment process. Rather than being buried in the lag coefficients, the effects of expectations are reflected by the Q variable in the investment equation.

Parallel to the theoretical debate about the appropriate specification of investment equations there is an equally established literature on the empirical evaluation of the incentive, allocative, and distributive effects of taxation. While taxation is usually an important item on the policy agenda, it is now receiving special attention. In part, this increased scrutiny has been caused by the recent tax reform in the United States which has a tremendous influence on the tax reform plans in Western Europe. Some papers in this volume contribute to this discussion, especially with respect to Germany and Great Britain.

The first paper of the volume by Pierre Poret and Raymond Torres provides an empirical cross-country analysis of various investment hypotheses. In order to compare the performance of standard investment equations for five countries (United States, Japan, Germany, France, Italy) two basic specifications are considered: a Keynesian flexible accelerator model and a profitability model. Finally, following standard practice these two models are combined into a mixed profit-accelerator model and this mixed model is then compared with a rival Q model. The Q variables typically enter the regression equations significantly, though the valuation ratio effect is weak in the United States. While these results provide some support for the connexion between investment and Q, J-type non-nested tests were not able to discriminate between the two models under consideration. Given the fact that the Q model is most helpful in assessing the impact of stock market movements on investment the estimated Q-type investment equations were finally used to simulate the possible effects of the October 19, 1987 stock market crash. To illustrate the dynamics of investment in the short run, the Tobin's Q equations have been implemented into the OECD INTERLINK model and a stock market drop by way of a reduction of the Q ratio relative to the baseline has been evaluated. Ignoring wealth effects of falling stock prices on consumption which have received some support from a number of recent econometric studies the impact on investment turns out to be small. This finding is in agreement with a growing literature that calls into question the "efficient markets" theory that all stock price movements must be interpretable by information about economic fundamentals.

Two papers, one by Hiroshi Yoshikawa and one by Stephen Bond and Michael Devereux, provide an analysis of investment behaviour taking explicitly into account the empirical problems of numerous econometric studies of investment using Q-type investment functions.

Hiroshi Yoshikawa takes a historical perspective on the postwar importance of fixed investment in Japan and then focuses on the fact that there is now a substantial body of direct evidence suggesting the importance of "excess volatility", "myopia" or other inefficiencies in stock market prices. These empirical results are certainly difficult to reconcile with simple Q models which assume that stock prices are closely related to investment. The purpose of the paper is therefore to provide a concise way to model investment behaviour in a situation where stock prices are "too volatile" relative to investment decisions. The author develops a model in which investment is a function of expected marginal Q and the three variables sales, real wages, and real energy costs upon which expectations of Q are conditioned respectively. Clearly, short-term fluctuations in the variables determining investment behaviour in the model will not be related to the investment decision of the firm and, for quarterly data, one needs some way to "filter-out" the noisy part of the variables from that part containing an accurate signal of the firm's prospects for the future (This problem will obviously weigh less heavily when using annual data). The suggested solution is to use the Beveridge-Nelson decomposition, a time series technique that is designed precisely as a filter and that can be used to separate "transitory" from "permanent" components of a time series. An alternative method would be to use more sophisticated econometric techniques like "band spectrum regression". The former procedure is used in the paper and the empirical tests, carried out by a series of multiple regressions using quarterly data from 1971Q1 to 1986Q2 for six Japanese manufacturing industries are on the whole in accord with the presented model. These results suggest that the distinction between transitory and permanent (fundamental) movements in asset prices may in fact enhance the performance of quarterly Q-type investment equations.

The starting point for Stephen Bond and Michael Devereux's paper is the fact that the empirical performance of Q-type investment equations is often regarded as mixed. Although Q typically enters regression equations for the rate of investment significantly, there is often the difficulty that the residuals from the investment equation are heavily serially correlated. This puzzling serial correlation which is inconsistent with the simplest form of the Q theory may be attributable to errors in measurement. The paper by Stephen Bond and Michael Devereux is a step towards answering this question and is part of an effort to consider this problem at the microeconomic level, i.e. using firm panel data instead of aggregate time series. They examine the degree to which these mixed results can be traced to measurement problems in the replacement value of the capital stock caused by the rapid increase in energy prices and subsequent obsolescence of part of the capital stock. They calculate alternative tax adjusted marginal Q values and investment rates using three considerable different estimates of the replacement value of the capital stock for an unbalanced panel of 191 quoted UK companies whose major activity is within manufacturing. The econometric evidence shows that in the levels specification both Q and the lagged dependent variable are significant, although the estimated coefficients on Q are low. Looking at the first difference specification instead of the level one, the coefficients on Q fall further and are insignificantly different from zero. Concentrating on the effect of the three different estimates of the capital stock, the pattern of results is very similar for all three capital stock and Q figures, i.e. the coefficients and their significance correspond to orders of magnitude which are roughly the same. The robustness of the results indicates that the problem of serial correlation does probably not stem from measurement problems in the replacement value of the capital stock. On the positive side, these results seem sufficiently compelling to warrant a careful examination of other possible sources of misspecification which may cause the mixed results of Q-type investment models in some data sets.

The paper by Michael Funke, Andreas Ryll, and Dirk Willenbockel adopts a Q-theoretic approach for the determination of investment expenditures in West German industries. The basic building block of the model assumes a competitive industry, whereas the industry is characterized by dynamically optimizing firms with rational expectations who take into account the complex tax structure in West Germany. The model therefore allows to establish a link between the determinants of investment and the role of tax policies in promoting economic growth. The theoretical model is consistent with estimation results for 17 West German industries over the sample period 1965-1982. In common with many other researchers, they conclude that measurable variants of marginal tax adjusted Q have high power in equations for fixed investment. In contrast with some other empirical work, however, they find no unequivocal evidence for an independent accelerator effect. This result is surprising at first sight since the stock market in West Germany plays a comparatively minor role in financing investment compared to the United States or Great Britain. What this could suggest is that the link between Q and investment arises less because of the direct part played by the stock markets than because the underlying sentiment that moves the stock market is also that which favours investment expenditures. To sum up, the results are lending weight to the argument that the valuation ratio should be taken into consideration in investment equations.

The second paper on Germany by Peter Trapp represents a distinctive approach to the problem of the secular decline in profitability. The paper re-examines the profits-investment link and falls neatly into the category of "causes, consequences, and cures". In order to gain a picture of the development of the rate of return on fixed assets and Tobin's Q the balance sheet and profit and loss account statistics of the Deutsche Bundesbank were used to calculate profitability measures. The calculations reveal the long-run decline of the rate of return in Germany. The policy issue of the paper is that according to Peter Trapp the downward trend in profitability stems not from a shortfall of aggregate demand but from unfavourable structural developments on the supply side of the economy. It is argued that the main reason for the deterioration of the rate of return is the increase in government interventions and regulations, particularly in the labour market, which have increased production costs and reduced the flexibility of the economy. In addition public assistance to declining sectors has contributed to real wage rigidity. Because an increase in the rate of return is seen as a precondition for future economic growth he calls for increased flexibility in wage determination, a change in employment protection legislation, and modification of social policy measures to reduce disincentives.

The paper of Elias Dinenis adopts a Q-theoretic approach for the determination of the effects of the 1984 Tax Reform on private nonresidential investment in the UK. In the UK, we have seen the phasing out and finally abolition of the first-year allowances, and significant reductions in corporate tax rates in three successive cuts. Consequently, it is not at all clear on an intuitive basis what impact the new law will have on investment decisions. Elias Dinenis' paper reports an effort to evaluate these impacts. The Q-theoretic methodology allows to capture the effects of the 1984 tax reform on Q and then using the estimated investment equation to calculate the short-run effects during the transition period as well as the long run impact of the tax reform on investment. The careful modelling of the tax effects shows that during the transition period the positive effect will dominate and investment will increase. The long-run effect on investment, however, is slightly negative amounting to a reduction of some 0.7 per cent.

For Willi Leibfritz the issue is the evaluation of tax policy measures actually taken in Germany and how they may alter investment spending. He first considers the tax rates that apply to incremental investment decisions using the well known methodology described by King and Fullerton. They measure the real effective marginal tax rate on additions to the capital stock of the domestic non-financial sector that are financed out of savings of domestic households. In the second part of the paper he additionally applies the user cost of capital approach originally defined by Hall and Jorgenson. For both concepts updated calculations for Germany are presented. Furthermore a projection is shown until 1990 when the tax reform will be implemented. This projection is based on the "no change" assumption for non-tax factors in order to outline the isolated tax reform effect. The detailed figures show that the effects of the 1990 tax reform package will only be moderate. Measured with the traditional user cost of capital approach capital cost will increase slightly while the integrated tax model of King/Fullerton shows a moderate decline in the marginal tax rates. These calculations are supplemented by the results of business surveys carried out by the Ifo Institute to assess the impact of taxes and capital costs on business investment. In these questionnaire surveys taxation does not appear to be a major factor in firms' investment decisions, however, it does emerge as a subordinate element in a complex decision situation. The survey reports indicate that sales and profitability are the most important factors determining the level of investment. The major expansionary effect of the 1990 tax reform package in Germany will therefore probably result from the reduction of the personal income tax helping to stabilize overall demand.

The emphases the conference has put on cross-country empirical analyses is clearly reflected in Julien Alworth and Wilhelm Fritz's paper which determines the appropriate measurement of effective tax rates in an open economy. What is the innovative aspect of the paper is that the authors examine how the effective tax rates obtained when using the King/Fullerton framework need to be modified in an open economy where domestic investments and savings need no longer be equal. In the first part of the paper the issues involved in defining a financial equilibrium in an open economy where interest rates are determined in the world capital markets are set out. The specification of an equilibrium is then used in the second part of the paper to calculate effective tax rates in the presence of an open economy for a selected group of nine OECD countries. To keep the burden manageable, the study is limited to the effective tax rates within each single country as in King/Fullerton, and not for the full range of possible combinations of countries and investors. Effective tax rates for nine countries are computed for the same investments as in King/Fullerton with the major effect of "openness" being that of altering the value of interest rates faced by domestic savers and investors. The general impression from the simulation results is one of considerable convergence in the effective tax rates, i.e. the results suggest that the existence of capital mobility alters the value of effective tax rates in a significant manner. The assumption of closed economies may therefore have seriously biased upward the actual extent of cross-country differences in the effective tax rates.

The paper of Gordon Anderson in this volume provides a Canadian perspective and estimates the tax effects on corporate investment expenditures and the corporate financial structure. Elaborating on earlier work by the author, the paper gives a careful modelling of the tax status of individual firms and estimates the tax effects in a microeconometric study. The result is that the tax policy incentive effects for investment expenditures are far more reaching and significant than has been indicated by aggregated empirical research thus far. He suggests that more attention should be paid to microeconometric studies which allow a precise evaluation of the complex tax status of individual firms. These could well be lost in the aggregation process concomitant with macroeconomic analysis.

The paper of Toshiaki Tachibanaki is devoted to the financing of business investment. It addresses two sets of questions. First, how have the financial systems in the four countries under consideration (Japan, United States, Germany, United Kingdom) performed in funding business investment? Second, to what extent can the special pattern of financing in Japan be explained by taxation or other institutional arrangements? The first section of the paper contrasts the interaction between the supply for investment funds and the demand for investment funds in the four countries, i.e. attention is paid to the sources of finance of investment. The four countries can be classified into two groups based on the structure of the financial markets, namely Japan and West Germany, versus the United States and the United Kingdom. The picture shows that the share of bank loans in financing physical investment is considerable high in Japan and Germany ("indirect financing"), while the roles of internal funds are more important in the United States and in the United Kingdom ("direct financing"). To what factors can this "loan-based system" in Japan be attributed? The paper gives evidence that the tax system in Japan is very favourable of private deposits at banks and postal offices. At the same time the "main bank system" leads to very close relationships between banks and their customers. The overall effect of this institutional fact is that Japanese banks provide to some extent the functions that one would otherwise expect of a risk sharer, namely equity type services. In other words: The Japanese "loan-based system" with its high proportion of debt in financing investment shows most of the characteristics of equity finance. The author suggests that this feature of the financial system contributes to the high investment rates in Japan.

The paper by Karl Aiginger follows a theoretical bent in considering the investment behaviour of business firms in an uncertain environment. The central question in the first part of the paper is under what conditions uncertainty theory provides an unambiguous answer to the problem whether firms facing a situation of uncertainty will produce and invest more, the same, or less than under certainty. In presenting a short survey of models of decisions of firms based on the method of expected utility maximizing behaviour, the paper derives four general propositions describing the impact of uncertainty depending on risk attitude, technological concavity, disequilibria and asymmetries in ex-ante and ex-post adjustment possibilities. The derived theoretical propositions result in a dichotomization of uncertain situations into those of "petty" and "severe" uncertainty, respectively. Situations in which ex-post adjustments are feasible are labelled as "petty" uncertainty, while situations in which there are less strategies for ex-post adjustment are labelled as "severe" uncertainty. The outcome is that uncertainty theory reveals a biasing down of production and/or investment in a situation of "severe" uncertainty. Knowledge of the ex-post adjustment possibilities is therefore important for an assessment of the impact from uncertainty.

While there are many common themes throughout the papers, there is also a diversity of opinions concerning the determinants of investment decisions. Readers of this volume should not therefore expect a new conventional wisdom to emerge. It is to be hoped, however, that the papers recorded here will stimulate a wider debate of, and further research into, this important subject.

What Does Tobin's Q add to Modelling of Investment Behaviour?

Pierre Poret and Raymond Torres

Economic and Statistics Department, OECD*

Introduction

Although many researchers have drawn attention to the theoretical relevance of Tobin's Q as a possible financial determinant of investment behaviour, this has had little or no impact on macro-economic modelling and most of the large models still include traditional investment equations. The stock market crisis in October 1987 has clearly illustrated the need for a larger integration of financial and real sides of the macro modelling. This paper seeks to make a modest contribution to this issue by comparing the performance of standard investment equations for five OECD countries with those including a measure of Tobin's Q.

---

* Views expressed are those of the authors, and do not necessarily reflect those of the Organisation. Helpful comments were received from John P. Martin.

Part I presents an internationally comparable method of calculation of Tobin's Q in the business sector for the United States, Japan, Germany, France and Italy (1), in line with the approach followed for France in Chan-Lee (1986). Part II describes theorical advantages and practical caveats of the Q ratio. Part III compares Q and standard models of investment. It concludes that, while in some countries the Q model tracks slightly better the short-term consequences of shocks, it embodies too many speculative fluctuations leading to a deterioration in its explanatory power on average over the sample period. Then the possible effects of the recent fall in stock prices have been simulated by Interlink - the OECD macro model - in which the estimated Tobin's Q equations have been implemented, by way of experiment.

I. Definition and measurement of Tobin's Q

Tobin's Q is defined as the ratio of the market value of the representative firm to its capital stock, valued at its replacement cost. When such a ratio exceeds unity, it is profitable for the firm to invest in new capital goods, since the expected return to such an investment will exceed its cost. However, while a number of measures of the capital stock are available for a large number of OECD countries, this is not the case for the so-called "market value of the aggregate firm". A methodology for its calculation is briefly developed below.

The balance sheet of the aggregate firm is assumed to consist of equity and debt. It is assumed that the financial market valuation of equity is equal to the present value of the sum of future dividend payments. If one assumes that the discount rate of future dividend payments is equal to the dividend yield -- as would be the case if the market efficiency hypothesis holds -- a very simple measure of the market value of equity can be derived by dividing the value of current dividend payments by the current dividend yield: FP = div/d, where div represents the net dividend payments, d the current dividend yield and FP the market value of equity. Implicit in this calculation is the assumption that the expected future dividend payments are constant over time. This static expectation hypothesis is technically unavoidable, but implies that our Q measure does not fully incorporate all relevant information as suggested by the Tobin's Q theory.

Similarly, the market value of the debt component of the balance sheet of the representative firm is assumed to be equal to the ratio of current interest payments to some current long-term interest rate: D = int/R where int represents the net interest payments, R the long-term interest rate and D the market value of debt. Once again, a condition for ensuring that the identity holds is that the flows of expected interest payments are constant over time.

The average Tobin's Q ratio therefore is (D + FP)/K. The existence of different tax burdens on interest and dividend revenues is not considered in this study. Although the proposed definition of Tobin's Q is a very simple one, it includes both the cost of equity and the cost of debt and, in that sense, is more general than other definitions of Q. Some authors [for instance, Sachs et al. (1985)] only consider the ratio of share prices to the investment deflator, and therefore ignores the cost of debt. On the other hand, in a number of studies the investment behaviour is only explained by deviations of the rate of return on capital from the real rate of interest. Table 1 presents the values of Tobin's Q for the five OECD countries under study. Given that R and d are not tax-adjusted, that self-financing is not taken into consideration and that d only concerns large companies quoted on stock markets which are probably more profitable than the average, the numerator (D + FP) is biased downwards. As a result, the Q ratio never exceeds unity and has been normalised. Hence, only the international comparison of the Q patterns presents some interest. In Table 1, two important aspects emerge. First, Tobin's Q is very volatile, in line with the financial market expectations reflected in R and d. Second, this profitability indicator declined in all countries except Japan after the first oil shock. The period since the second oil shock, especially since 1982, has seen some recovery in Tobin's Q, although the pre-1973 levels were only re-attained in Germany and to some extent in France.

Table 1. TOBIN'S Q IN FIVE OECD COUNTRIES

(1975 = 1.00)

	United States	Japan	Germany	France	Italy
1971	1.472	1.152	1.022	1.316	1.067
1972	1.582	1.426	1.076	1.397	1.165
1973	1.539	1.276	1.036	1.340	1.056
1974	1.174	0.917	0.934	1.178	0.919
1975	1.000	1.000	1.000	1.000	1.000
1976	1.127	1.050	0.939	0.918	0.855
1977	1.052	1.105	1.278	0.850	0.774
1978	0.951	1.178	1.277	0.904	0.874
1979	0.924	1.030	1.194	1.015	0.797
1980	0.913	1.000	1.039	0.856	0.891
1981	0.899	1.019	0.837	0.739	0.899
1982	0.819	0.945	1.202	0.771	0.789
1983	1.035	1.064	1.378	0.894	0.864
1984	1.071	1.167	1.348	0.989	0.921
1985	1.109	1.190	1.595	1.237	0.930

Source: National accounts and OECD Financial statistics

II. Theoretical advantages and practical caveats of Tobin's Q for the empirical modelling of investment behaviour

In principle, a key advantage of Tobin's Q is that it reflects all constraints on physical capital growth, including insufficient tax-adjusted production profitability, weakness of the demand for goods or rationing of the variable production factors. The Q ratio embodies directly the implicit weighting system for these various regimes, used by financial agents. In contrast, because of their linearity, the standard econometric models of investment generally assume that the relative importance of each regime is constant (2). Moreover, they make simplistic assumptions when specifying the main constraints on investment: in particular, unobservable demand constraints representing the Keynesian regime are proxied by firms' output or sales which may in fact be influenced directly by profitability constraints as well.

Nevertheless, while the relevant concept for investment decisions is the marginal value of Q, only the average value of Q is directly measurable from financial indicators. We know (Hayashi, 1982) that average and marginal values of Q are identical only if there is full market clearing and if returns to scale are constant. For instance, in excess supply, i.e. in a situation where profits do not reach their maximum levels due to demand constraints on production, the average value of Q will exceed its marginal value and a loosening of the constraint will reduce the gap between the two [see Malgrange and Villa (1984)]. This is one possible rationale for adding a demand pressure variable such as the capacity utilisation rate in the econometric specification of the Q model, a practice which often yields better results. Thus, in practice, Q models are less general than the Q theory would let the econometricians hope.

Another theoretical advantage of Tobin's Q is that it takes account of financial market expectations and agent's attitude towards risk. Only ex post profitability can be taken into account in the standard models; anticipated profitability is proxied by some arbitrary distribution of lagged profitability variables, usually according to an adaptive expectations mechanism. In principle, Q models do not involve such a restriction and there is no need to impose a rigid distribution on lagged profitability (or any other a priori solution). In practice, the addition of a lagged dependent variable on the right-hand side generally improves the statistical quality of regressions on the Q model. Although these lags are justified by the existence of delivery lags, they are actually intended to neutralise the excessive volatility of the discount rate entering the definition of Tobin's Q. This discount rate is precisely the component left to the subjective judgment of financial markets. On the contrary, the current profit rate, the other component of Q, is generally highly significant and has often more explanatory power than the discount rate in investment equations [see Ueda and Yoshikawa (1986), Abel and Blanchard (1986)]. Two different explanations have been suggested for the lack of correlation between the discount rate and investment. First, changes in financial assets take place much faster than changes in physical assets. This is because of a much greater degree of reversibility of financial transactions (existence of a secondary market), so that the market value of shares is very sensitive to any change in uncertainty and to new information. Second, stock markets in a number of countries are very narrow, and bank loans in these countries are the main source of external financing for domestic firms. In this context, credit is often rationed and the interest rate may not reflect the equilibrium discount rate. This was particularly the case in France and Italy, before the very recent wave of financial innovations and deregulation.

Finally, from the point of view of empirical modelling, it is difficult to implement Tobin's Q into a complete macro-model of the economy because some relevant variables (political climate, uncertainty) necessary to forecast the Q ratio are generally missing. Unpublished work by the OECD Secretariat has shown that variables such as the rate of return on capital, the inflation rate and the nominal interest rate only explain a small part of the variability in Q across countries and over time.

III. Comparative performance of Tobin's Q and traditional investment equations

Two simple specifications of the so-called traditional models are considered: a Keynesian flexible accelerator model and a profitability auto-regressive model.

In the former, the optimal demand for capital (denoted by KBSTAR) is determined by solving the cost-minimization problem subject to (a) the constraint of supply being equal to or smaller than demand and (b) the technological characteristics of the production function. It is assumed that the production function is a nested, two-tier CES function with constant returns to scale. At the first level, substitution is possible between energy and a certain proportion of capital (the putty/semi-putty case) and at a second stage substitution is assumed to occur between the capital-energy aggregate and labour (3). Technical progress is assumed to be entirely labour enhancing. Empirical estimates of the technical progress term as well as the scale parameters and the elasticities of substitution of the production function are derived from the national sub-models of INTERLINK, the international macro-economic model of the OECD Secretariat. It will be noted that, contrary to usual practice, in this study demand is represented by total demand and not by production. The latter may result, a priori, either from demand or from the profitability level. Total demand is assumed to be less rationed by supply constraints than production could be because if, for some reason, domestic output is limited and cannot satisfy domestic demand, imports will make up the shortfall. Adjustment lags and the formation of expectations are specified in a conventional way by means of lagged dependent variables. The dependent variable is the growth rate of the capital stock (KBV) in the business sector. Results are reported in Table 2. All equations are estimated on semi-annual data over the period 1971 SI to 1985 SII.

Table 2

Accelerator model: $\Delta \ln KBV = a + b \ \Delta \ln KBV(-1) + c \ \Delta \ln KBSTAR$ OLSQ: S1 1971 - S2 1985

(t- statistics in parenthesis)

	a	b	c	SEE	DW	$\bar{h}$	R^2	SEE with the constraint c = 1 - b
United States	$-0.38.10^{-2}$ (-2.5)	1.168 (14.2)	0.055 (7.3)	$0.945.10^{-3}$	1.50	1.25	0.8739	$0.103.10^{-2}$
Japan	$-0.26.10^{-2}$ (1.8)	0.893 (22.9)	0.005 (0.2)	$0.253.10^{-2}$	2.00	-0.08	0.9600	$0.280.10^{-2}$
Germany	$0.15.10^{-2}$ (2.3)	0.865 (23.7)	0.029 (3.2)	$0.990.10^{-3}$	2.00	-0.60	0.9560	$0.113.10^{-2}$
France	$0.37.10^{-3}$ (0.6)	0.909 (31.2)	0.054 (6.2)	$0.770.10^{-3}$	1.59	0.99	0.9790	$0.780.10^{-3}$
Italy	$0.14.10^{-3}$ (0.1)	0.925 (16.3)	0.035 (3.5)	$0.150.10^{-2}$	1.27	1.86	0.9000	$0.150.10^{-2}$

In the profitability model (Table 3), firms are assumed not to face demand constraints but the growth rate of their capital stock is assumed to depend upon the difference between profitability and the expected real rate of interest on financial investments (IRLRE, which is proxied by a moving average of real long-term interest rates). In fact, in all countries except France, IRLRE is hardly significant. Its degree of statistical significance depends on the choice of the profitability indicator: except for Italy, the rate of return on capital (TREN) defined as the ratio of gross operating surplus (i.e. before payments of interest) to the value of the capital stock gives better results than the profit rate (TPRO, which represents profitability net of financial costs). This is probably due to the fact that the interest rate effect is partly taken into account in the TPRO term. There is indeed a correlation between TPRO and IRLRE because in periods of high inflation (marked disinflation) both the relative weight of financial costs and the real interest rate decrease (increase). In the absence of strong and significant real interest rate effects, the profitability model therefore is quite similar to a cash-flow model of investment.

Table 3

Profitability Model: $\Delta \ln KBV = a + b \ \Delta \ln KBV(-1) + c \ TREN + d \ IRLRE$ OLSQ : S1 1971 - S2 1985

(t- statistics in parenthesis)

	a	b	c	d	SEE	DW	h	$\bar{R}2$
United States	-0.014 (-4.7)	0.655 (8.9)	0.116 (7.2)	$-0.7.10^{-2}$ (-0.2)	$0.96.10^{-3}$	0.983	2.77	0.869
Japan	$-0.27.10^{-2}$ (-1.45)	0.651 (9.7)	0.141 (3.95)	-0.089 (-1.4)	$0.202.10^{-2}$	2.63	-1.87	0.972
Germany	$-0.72.10^{-2}$ (-1.8)	0.820 (21.0)	0.095 (3.7)	-0.086 (-1.2)	$0.952.10^{-3}$	2.20	-1.1	0.960
France	$-0.61.10^{-2}$ (-3.54)	0.699 (11.5)	0.111 (5.3)	-0.170 (-3.0)	$0.848.10^{-3}$	1.62	1.06	0.975
Italy	$-0.3.10^{-2}$ (-2.0)	0.709 (7.8)	0.093 (3.8)	-0.0157 (-0.3)	$0.149.10^{-2}$	1.21	2.43	0.903
(with TREN replaced by TPRO)	$0.67.10^{-2}$ (4.6)	0.714 (8.6)	0.0506 (4.6)	-0.117 (-1.9)	$.1372.10^{-2}$	1.39	1.66	0.918

Chart 1 in annex reports actual and predicted capital stock growth rates for the two standard models. It is difficult to discriminate between them. In general, both over-estimate the capital growth when it is decelerating and conversely under-estimate it when it is accelerating. Some noticeable exceptions are found in the USA in 1974 SI, France in 1975 SI and 1976 SI and Italy in 1985, where residuals show opposite signs. The lags in the profitability model are shorter than in the accelerator model and the latter presents less error autocorrelation than the former. In line with the usual practice in macro-modelling of investment, these two standard models are combined into a mixed profit-accelerator model. The mixed model is then compared with the Q model. As its statistical properties are quite close to those reported for the profit accelerator models, only its SEE and h-autocorrelation statistics are reported for comparison in Table 4.

Table 4

Tobin's Q model: $\Delta lnKBV = a+b\ \Delta lnKBV(-1)+cQ+dIFU$ OLSQ:S11971-S21985

(t- statistics in parenthesis)

	a	b(1)	c(2)	d(3)	SEE	DW	H	$\bar{R}^2$	Memo item: "mixed" model SEE	Memo item: "mixed" model H
United States	-0.035 (6.70)	0.629 (9.34)	0.0074 (1.81)	0.039 (5.67)	$0.74.10^{-3}$	1.79	0.30	0.922	$0.89.10^{-3}$	2.92
Japan	-0.049 (1.87)	0.666 (3.60)	0.0091 (2.03)	0.045 (1.67)	$2.23.10^{-3}$	2.31	n.a.	0.966	$2.06.10^{-3}$	-1.87
Germany	-0.018 (2.40)	0.894 (18.45)	0.0064 (2.40)	0.016 (2.14)	$1.02.10^{-3}$	2.04	-0.54	0.954	$0.96.10^{-3}$	-1.13
France	-0.037 (3.71)	0.668 (8.44)	0.0095 (3.49)	0.041 (3.58)	$0.91.10^{-3}$	1.48	1.54	0.971	0.7010^{-3}	0.81
Italy	-0.023 (3.64)	0.801 (12.47)	0.0453 (3.30)	0.018 (3.10)	$0.98.10^{-3}$	1.75	0.06	0.958	$1.20.10^{-3}$	1.27

(1) In the equation for Japan, the two-periods lagged dependent variable also appears, with a coefficient of 0.200 (1.17).

(2) In the Italian equation, the second lagged Tobin's q, instead of its current value, is included. The United States' equation includes only first lagged Q. In addition, no stable and significant value for the coefficient of the Q term was uncovered in free estimation. Therefore, an interactive dummy variable was added to the lagged Q variable, which implies a different coefficient before and after 1975. The coefficient presented in the table is the one relative to the 1975-85 period. The pre-1975 coefficient is equal to 0.0030 (0.984).

(3) For Italy an interactive dummy variable is attached to the IFU variable.

Both the average value of Tobin's Q and a capacity use term (IFU) are included on the right-hand side of the Q model equations. IFU is based on the INTERLINK supply-block [see Jarrett and Torres (1987)] and is defined as the ratio of actual output to potential output. As suggested above, this capacity use term is added to capture any possible difference between the marginal Tobin's Q and the average Tobin's Q. The regression results shown in Table 4 indicate significant Q and IFU effects. While in the Japanese, German and French equations, the Q terms are nearly as significant as the IFU term, this is not the case for Italy and especially for the United States. In the Italian equation a stable parameter could only be found for the capacity use term when an interactive dummy variable was added to the equation. In the United States (ironically the country with the largest stock market), the Q effect is insignificant; it did, however, become significant after 1975 when an interactive dummy variable was added. In both countries, without the dummy variables no significant estimates could be found.

The SEE of the Q model is better in the case of the United States and Italy only, but for both countries interactive dummy variables were necessary. However, for all countries, Chart 2 in annex shows that both the mixed and Q models produce similar patterns of residuals. This suggests that, on average, the Tobin's Q model does not add any new information to the traditional one. In order to verify this, a J-type non-nested test [see Davidson and McKinnon (1981)] has been undertaken. Although the two models are supposed to explain investment, neither can be derived from the other one by way of restrictions on parameters. The two models being rival, a non-nested test is required. The hypothesis of non-rejection of the Q model by the mixed model is tested by adding the predicted values of the Q model - Qhat - to the set of mixed model explanatory variables. If the t-statistic attached to the coefficient of Qhat is significant, the hypothesis is accepted: the mixed model cannot reject its rival, the Q model. In the first row of Table 5, the results of this test are presented. Only in the case of the United States and Italy, both countries where interactive dummy variables were necessary, are the Qhat predicted values significant at a 5.p.c. level. For these two countries, the Q model is not rejected by the mixed model, and therefore contains specific information. For the other three countries, the Q model is rejected by the mixed model. The non-rejection of the mixed model by the Q model is reported in the second row of Table 5. In all cases, the coefficient of the predicted values of the mixed model is significant. This implies that the mixed model contains at least as much explanatory power as does the Q model. For the United States and Italy, both models contain different pieces of information and, in that sense, are complementary. In the case of Japan, Germany and France, the tests therefore suggest that the Tobin's Q model does not add any information to the mixed accelerator-profitability model and is therefore inferior to it.

Table 5. NON-NESTED TESTS (J TESTS)

Hypothesis	United States	Japan	Germany	France	Italy
The mixed model cannot reject its rival, the Q model	accepted	rejected	rejected	rejected	accepted
The Q model cannot reject its rival, the mixed model	accepted	accepted	accepted	accepted	accepted

Nevertheless, a major difference between the two models concerns the first-order autocorrelation statistics. In Table 4, these statistics indicate that residuals of the mixed model are not generated by a white noise process -- except in France. This is certainly because of the adaptive expectations mechanism adopted in the traditional models. While in the absence of shocks the adaptive expectations hypothesis may not be rejected, this is no longer the case in the presence of shocks. The Q model does not share this caveat with the traditional models and tends to track better at the turning points.

In order to illustrate their dynamics in the short run, the Tobin's Q equations for the four major OECD countries have been implemented into the INTERLINK model and a "stock price shock" has been administered by way of a reduction of the Q ratio relative to baseline levels (4). The baseline is a preliminary unpublished short-term forecast prepared for the OECD Economic Outlook of December 1987 and not yet taking account of the stock market crash at that time. Values for Q have been produced until 1989 SII by inverting the investment equations (5). For the first year (1987 SII-1988 SI), the Q ratio has arbitrarily been reduced by 10 per cent. This is less than the current fall in stock prices, but it must be remembered first that the market value of firm's assets does not only consist of equities and, second, that not all firms are quoted on the stock market. After the first year, it is assumed that the Q ratio gradually recovers to its baseline levels for two reasons. Firstly the capital stock adjusts. Secondly the debt/equity ratio is endogeneous and should react to changes in the relative costs of shares and loans in order to minimize the overall cost of capital. Therefore, for the second year (1988 SII-1989 SI) a reduction of 5 per cent has been imposed and finally Q is assumed to revert to its baseline level for the last semester (1989 SII). In Table 6, simulation results for investment and capital stock growth are given. Readers must be aware that the wealth effects coming from other demand components are ignored (e.g. the effects of falling stock prices on household's wealth and consumption have not been considered) but that multipliers have negative effects on business investment.

Table 6. RATES OF GROWTH OF BUSINESS SECTOR INVESTMENT AND CAPITAL STOCK

		1987:I	1987:II	1988:I	1988:II	1989:I	1989:II*
United States							
KBV	Baseline	3.17	3.24	3.28	3.28	3.27	3.26
	Simulation	3.17	3.24	3.19	3.14	3.14	3.14
IBV	Baseline	-3.80	5.20	4.20	3.30	3.10	2.90
	Simulation	-3.80	5.20	1.80	1.60	3.20	3.10
Japan							
KBV	Baseline	6.38	6.28	6.21	6.16	6.12	6.10
	Simulation	6.38	6.09	5.84	5.71	5.62	5.65
IBV	Baseline	8.61	5.00	5.30	5.60	5.80	6.00
	Simulation	8.61	0.94	1.20	3.73	4.02	6.64
Germany							
KBV	Baseline	2.98	3.02	2.93	2.97	2.91	2.92
	Simulation	2.98	2.93	2.77	2.77	2.70	2.74
IBV	Baseline	1.95	4.45	-0.24	4.17	1.06	3.28
	Simulation	1.95	1.22	-2.88	2.94	0.06	4.08
France							
KBV	Baseline	2.15	2.26	2.29	2.34	2.38	2.42
	Simulation	2.15	2.17	2.11	2.14	2.16	2.23
IBV	Baseline	1.80	5.90	3.00	4.10	3.70	3.50
	Simulation	1.80	2.71	0.41	3.09	2.81	4.27

Notation: KBV, gross capital stock and IBV, for investment in the business sector.
* Investment growth in 1989:II is higher in the simulation than in baseline for two reasons. First, simulated capital growth starts increasing in that semester. Second, it does so from a low level of simulated investment. Of course, at the end of the period the simulated level of investment is still below baseline.

Concluding remarks

In none of the five countries does an investment equation incorporating Tobin's Q explain the investment behaviour at the aggregate level better than traditional models do.

This conclusion basically results from the method chosen for measuring the Q ratio and does not imply that Tobin's Q theory is irrelevant. While only the average Tobin's Q is measured, the relevant concept is the marginal Q; although some attempts have been made to capture the specific effects of marginal Q, more research is still required. Also, its numerator is built on the assumption that the firms' financing structure is neutral with respect to investment decisions and no attempt has been made here to split and estimate the effect on investment behaviour of each component of the firms' market value.

However, comparisons made between Q models and standards models point out that the main weakness of the latter arises from their adaptive expectations nature. Based on the financial expectations of agents, the Q model avoids systematic lags in the adjustment of expectations to new economic conditions and therefore is likely to eliminate an important source of residual autocorrelation. This illustrates the need for implementing alternative expectations formation mechanisms in the macro-models and/or for finding solutions to the residual autocorrelation problem in the context of adaptive expectations. In addition, while the discount rate effect on investment does not prove statistically significant in the standard profitability model, the Q model provides a clear transmission from financial markets to the real side of the economy. Altogether the Q model appears to be the best one for trying to quantify the possible impact of the stock market crisis on investment. even if this impact turns out to be small.

NOTES

1. United Kingdom and Canada are excluded because national appropriation accounts distinguishing dividend and interest payments for the business sector are not available.

2. However, more recent empirical studies, following a disequilibrium framework, have tested less constrained models, allowing for moving weights of the various regimes by calculating them either directly from business survey data (see Lambert and Mulkay, 1986), or from simultaneous estimates of regime probabilities (see Poret, 1986).

3. For an exhaustive description of this particular production function, see the article by Helliwell et al. (1986).

4. The shocks have been produced in unlinked mode, with constant money supply and constant nominal public expenditure.

5. Note that the Q value includes the residual term of the investment equation. Because the simulations are baseline-dependent. our results are therefore somewhat inaccurate, although we do not believe this problem is a serious one.

BIBLIOGRAPHY

Abel, A.B. and Blanchard, O.J.: "The Present Value of Profit and Cyclical Movements in Investment", Econometrica, Vol. 54,2, March 1986.

Chan-Lee, J.: "Pure Profits and Tobin's Q in Nine OECD Countries", OECD Economic Studies, N° 7, Autumn 1986.

Chan-Lee, J. and Torres, R.: "Q de Tobin et taux d'accumulation en France", Annales d'Economie et de Statistique, 5, 1987.

Davidson, R. and McKinnon, J.G., "Several Tests for Model Specification in the Presence of Alternative Hypothesis", Econometrica, 49, 1981.

Hayashi, F.: "Tobin's Marginal Q and Average Q: A Neo-Classical Interpretation", Econometrica, January, 1982.

Helliwell, J.F., Sturm, P., Jarrett, P. and Salou, G.: "The Supply Side in the OECD's Macroeconomic Model", OECD Economic Studies, No. 6, Spring, 1986.

Jarrett, P. and Torres, R., "A Revised Supply Block for the Major Seven Countries in INTERLINK", OECD Economic and Statistics Department Working Paper N°. 41, April 1987.

Lambert, J.P. and Mulkay, B.: "Investment in a Disequilibrium Context or Does Profitability Really Matter?", CORE Working Paper, October 1986.

Malgrange, P. and Villa, P.: "Comportement d'investissement avec coûts d'ajustement et contraintes quantitatives", Annales de l'INSEE, 53, 1984.

Poret, P.: "Modèles de demande de capital en déséquilibre et enquêtes de conjoncture dans l'industrie française", Recherches Economiques de Louvain, 1988 (forthcoming).

Sachs, J. and Wyplosz, C.: "Mitterand's Economic Policies", INSEAD, 11-12 November, 1985, unpublished mimeo.

Ueda, K. and Yoshikawa, H.: "Financial Volatility and the Q Theory of Investment", Economica, Vol. 53, February 1986.

ANNEXES

CHART 1. PERFORMANCE OF TRADITIONAL MODELS

———— Observed capital growth rate
- - - - - Fitted profitability model
—— – — Fitted accelerator model

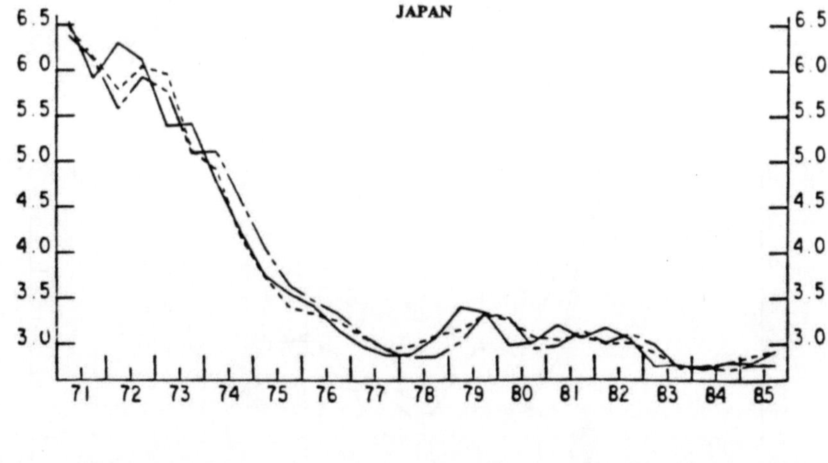

Chart 1 (continued)

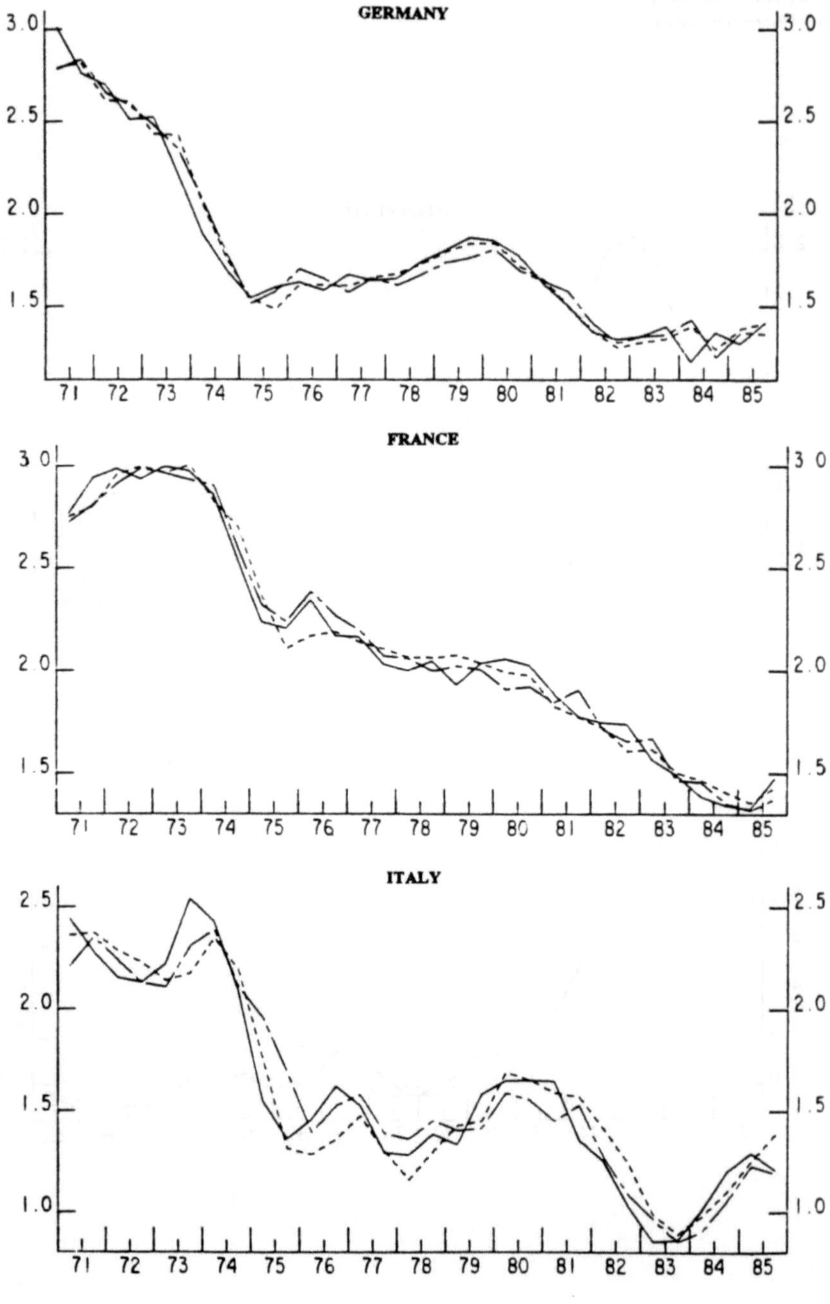

CHART 2. COMPARATIVE PERFORMANCE OF TRADITIONAL AND TOBIN'S Q MODELS

CAPITAL GROWTH - USA

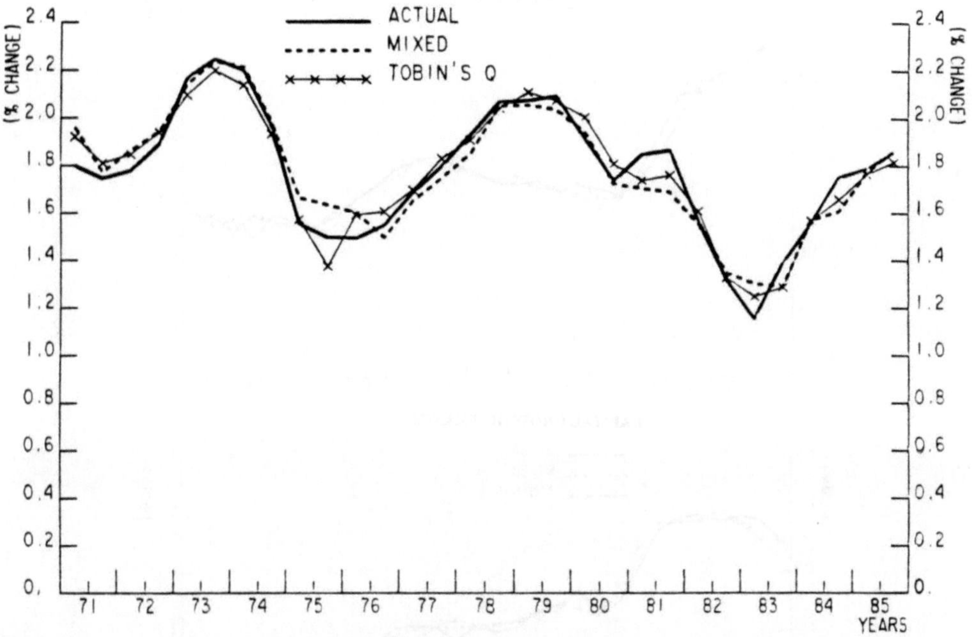

CAPITAL GROWTH - JAPAN

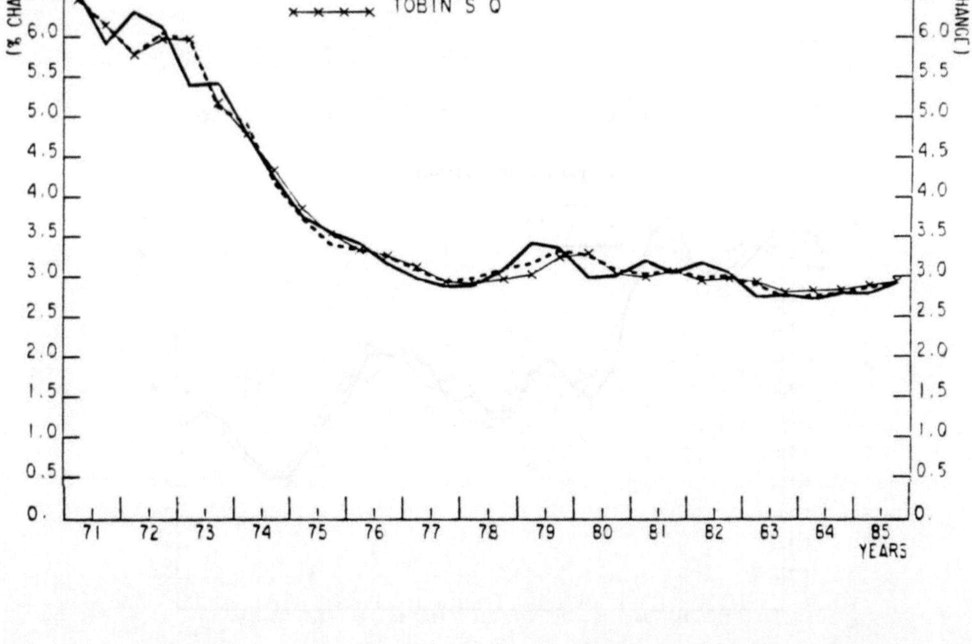

Chart 2 (continued)

CAPITAL GROWTH - GERMANY

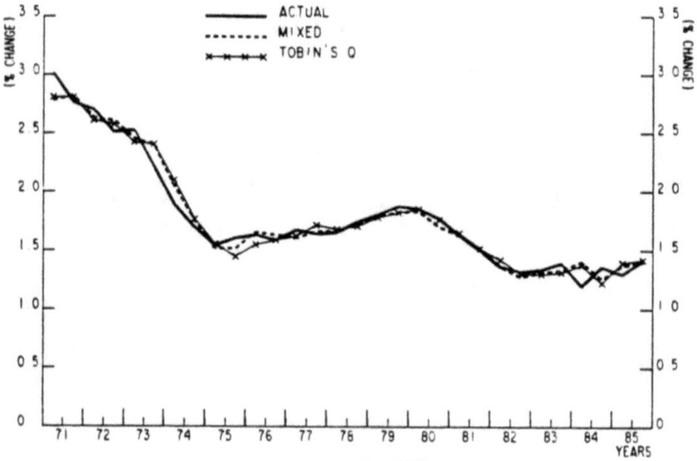

CAPITAL GROWTH - FRANCE

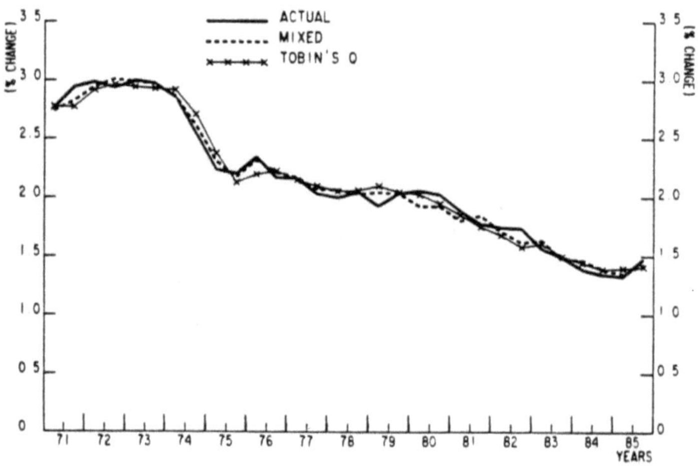

CAPITAL GROWTH - ITALY

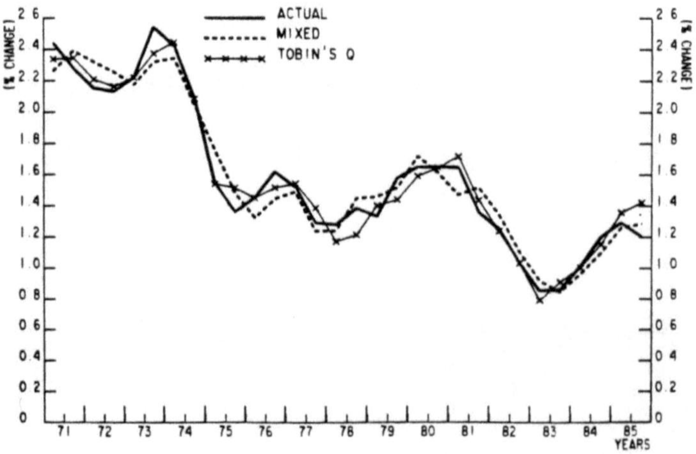

Fixed Investment in Japan

H. Yoshikawa

Institute of Social and Economic Research,
Osaka University, Ibaraki, Osaka, Japan

This paper studies fixed investment in Japan. The most notable characteristic of the postwar Japanese economy was perhaps its very rapid economic growth, a trait it shared with West Germany. Investment not only determines the rate of capital accumulation, but to the extent that most modern technological innovations are embodied in new capital goods, it also brings technical progress. Through these two channels, it plays a decisive role in determining the rate of economic growth. In the short–run, investment is a very volatile component of aggregate demand, and therefore greatly affects macroeconomic fluctuations. Clearly it is essential for us to understand more profoundly the behavior of investment in order to better grasp the movements of the macroeconomy both in the short–run and the long–run.

A standard method of studying investment is, of course, estimating econometric equations. Before proceeding to the estimations of such equations, however, it is very useful to consider the behavior of investment in a historical context. After all, we could not hope that a particular theory would uniformly apply to all countries, nor would we expect it to hold over all time periods. We therefore suggest that analyses of investment in various countries will prove especially useful in this regard.

This paper consists of three sections. In Section 1, we briefly summarize some important stylized facts concerning fixed investment in Japan. It is a macroeconomic and also historical overview. In Section 2, we ask ourselves what are the important determinants of investment. Here we propose a theory of investment which emphasizes the distinction between permanent and trasitory shocks in the determinants. Using the quarterly data of Japanese manufacturing industries, we estimate this investment function for the post–first oil shock period (1972. II – 86. III). Section 3 offers some concluding discussion.

1. Investment in Growth and Aggregate Fluctuations

Investment is nothing but a mirror image of economic growth. In Figure 1, we show rates of change in real GNP and corporate fixed investment in Japan. It is immediately clear that changes in investment are highly correlated with changes in real GNP, but are more volatile. On careful examination, it is also clear that the correlation between the two variables has got smaller to some extent over the last ten year sample period.

How important was the relative contribution of capital to high economic growth in Japan? This supply side problem is usually analyzed by growth accounting based on the neoclassical production function. In their recent study, Kuroda, Yoshioka and Shimizu (1987) obtained the results reproduced in Table 1. Their work is based on industry level growth accounting and therefore explicitly takes into account productivity growth due to shifts in the composition of industries. Table 1 underlines the significant role of capital in economic growth and also shows that the *relative* contribution of capital increased in the post–first oil shock period. Norsworthy and Malmquist (1983), comparing productivity growth in Japanese and U.S. manufacturing, also conclude that "the productivity 'miracle' in Japan is not so miraculous after all: It is largely explained by a higher rate of growth of capital and materials per worker."

There is no doubt that capital is the major contributor to economic growth. There still remains, however, a question of the role of investment in short–run economic fluctuations. In Figure 1, we have already noted the high correlation between changes in investment and real GNP. To proceed a step further, Yoshikawa and Ohtake (1987) carried out a Granger causality test for Japanese investment and real GNP. The result we obtained is that investment causes real GNP in the sense of Granger during the pre–first oil shock of rapid growth, but *not* during the post–first oil shock period. This finding suggests that Japanese business cycles in the 1950's and 60's were investment induced. It is interesting that this result of the causality test for Japanese data is in sharp contrast to the one for U.S. data. For the U.S., the result typically obtained is that investment is governed by real GNP, not the other way round. This is in accord with the commonly held view that in the U.S., residential construction and automobile purchases are the major generator of real GNP, which in turn induces corporate fixed investment: See also Gordon and Veitch (1986) in this regard.

Figure 1: Changes in Real GNP and Fixed Investment

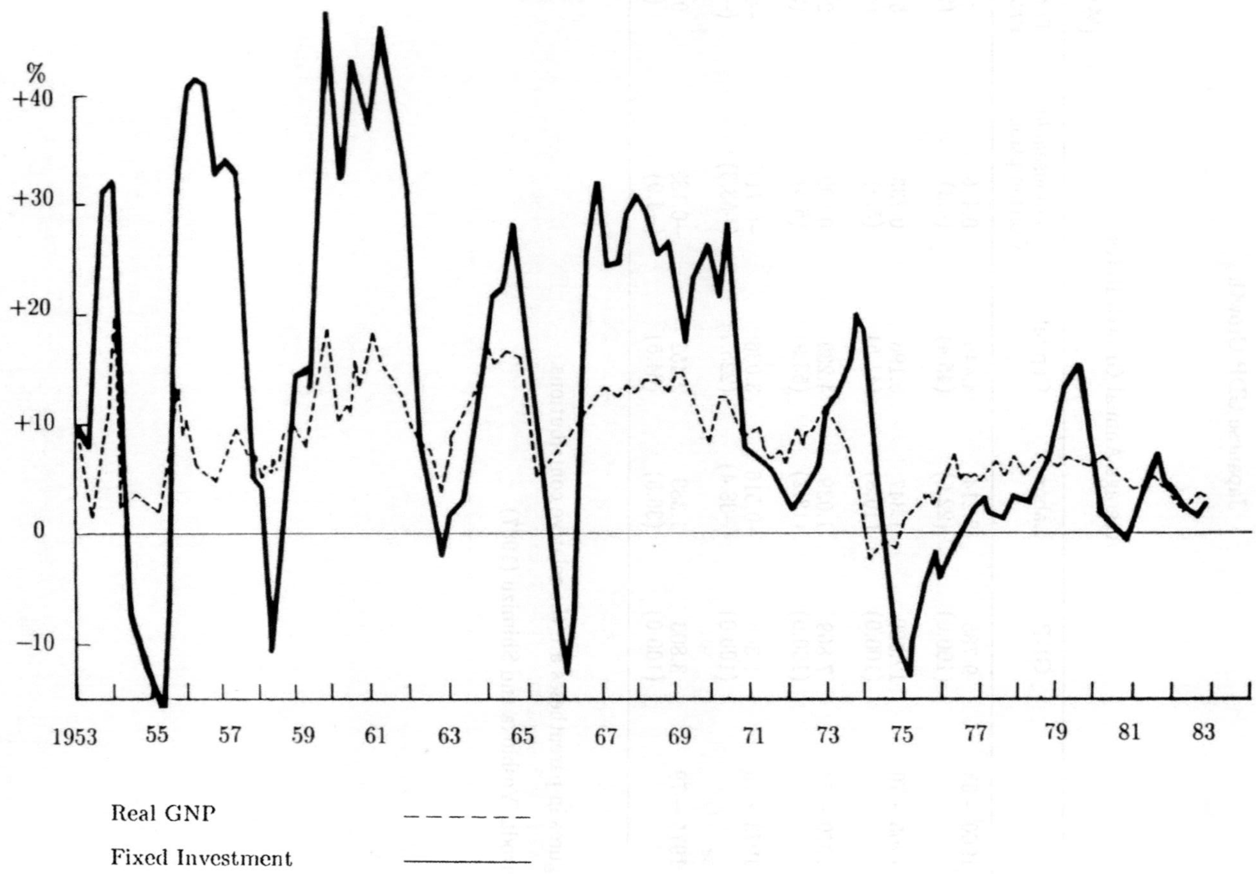

Table 1: Factor Decomposition of Japanese GDP Growth

Average Annual Growth Rates

(%)

	GDP	Labor	Capital	Nonhousehold Consumption	Total Factor Productivity
1960 – 65	9.785 (100.0)	2.214 (22.6)	4.440 (45.4)	0.416 (4.3)	2.714 (27.7)
1965 – 70	12.529 (100.0)	1.347 (10.8)	5.196 (41.5)	0.733 (5.9)	5.252 (41.9)
1970 – 73	7.869 (100.0)	1.039 (13.2)	4.239 (53.9)	0.509 (6.5)	2.081 (26.4)
1973 – 75	1.346 (100.0)	–0.516 (–38.4)	3.036 (225.7)	–0.717 (–53.3)	–0.458 (–34.0)
1975 – 79	3.803 (100.0)	1.389 (36.5)	2.452 (64.5)	–0.188 (–4.9)	0.147 (3.9)

Figures in parentheses are the relative contributions.

Kuroda, Yoshioka and Shimizu (1987).

We discussed some evidence suggesting that Japanese business cycles in the 1950's and 60's are investment induced. In the early 1970's, however, this pattern had changed. In Japan, it is widely believed that exports took place of investment in the 70's and 80's and they became the major contributor to aggregate fluctuations. For example, the percentage contribution of exports to the growth rate of real GNP was – 8.3% for the period 1963 – 73 whereas it was + 40.2% for the period 1980 – 84. Also the simple correlation between industrial output and exports was –0.41 for the period 1960. III – 71. IV whereas it was + 0.56 for the period 1972. I – 85. IV. The change in the sign of correlation is consistent with the view that in the 60's changes in output were *not* export–led, perhaps investment–induced, and exports were rather a "vent for surplus production," whereas in the 70's exports have become a major force generating business cycles in Japan.

We also estimated a vector autoregression (VAR) model containing quarterly investment, exports, and domestic output to obtain the variance decompositions of each variable. The results shown in Table 2 suggest that investment was very autonomous in the period, 1966. IV – 73. I, whereas exports were very autonomous in the period, 1973. II – 84. IV. We also note that investment became very dependent on exports in the period, 1973. II – 84. IV.

In line with the change in the pattern of fluctuations, the average growth rates of both real GNP and investment sharply decelerated beginning the early 1970's: See Figure 1. It should be noted that this significant break in "trends" is not common for all the industrial countries. For example, a careful international comparison done by Horie, Naniwa and Ishihara (1987) shows that the sharp decline in trend of real GNP is not observed for the U.S. and U.K. This casts some doubt on the popular view (Bruno and Sachs (1985)) that the oil shock was the major cause of the break in "trends" observed for the Japanese economy.

Taking issue of this oil shock explanation, Yoshikawa and Ohtake (1987) emphasize that demographic trends played a particularly important role in the Japanese economy. The Japanese economy in the 1950's and 60's was a two–sector economy consisting of a rural agricultural sector and an urban manufacturing sector and population continuously flowed from the former into the latter. Because of these large–scale population flows, the number of households dramatically increased during this period. We argue that these population flows and the consequent increases

Table 2: Variance Decompositions of Investment, Export and Gross Domestic Demand (VAR with 4 Period Lags)

66. IV to 73. I (10 quarters ahead)

	Stand Error	Investment	Export	GDD
Investment	.0679	72.93	14.48	12.58
Export	.0603	29.35	39.91	30.74
GDD	.0339	51.96	20.18	27.86

73. II to 84. IV (10 quarters ahead)

	Stand Error	Investment	Export	GDD
Investment	.0494	43.28	51.53	5.18
Export	.0907	25.92	71.51	2.57
GDD	.0112	32.501	27.80	39.70

in the number of households continuously stimulated *economy–wide* investment. Note that for a wide range of goods such as consumer durables, the number of households is a more important determinant of demand than is population itself, and therefore not only residential investment but also plant construction is stimulated by a sharp increase in the number of households.

This situation changed dramatically around 1970. By then the pool of the so–called disguised unemployed in the agricultural sector had been largely exhausted. Therefore the population flow from the rural sector and the associated increase in the number of urban households both sharply decreased. We stress that this change had occurred well *before* the first oil shock in the fall of 1973. In any event, investment thereafter seems to be governed by micro–specific rather than economy–wide factors.

The point can be seen by looking at the correlations of investment across industries for both the rapid and low growth periods (See Yoshikawa and Ohtake (1987)). Then we observe strong parallels in investment across various sectors during the period of rapid economic growth. This suggests that economy–wide rather than industry–specific factors spurred investment demand in the 1960's. In contrast, in the 1970's and 80's, the correlations are low, very often negative, which suggests that investment in this period is largely governed by microeconomic rather than economy–wide factors. The important exception is the machinery block consisting of fabricated metals, general machinery, electrical machinery and transportation machinery (automobiles). This machinery block is very closely related to exports in the 70's and 80's. In fact, the relative share of this machinery block in total manufacturing investment has increased from 32.4% (1961) to 52.8% (1986). In contrast, the share of iron and steel has decreased by almost the same percentage.

We can summarize the above discussion as follows. The Japanese economy in the 1950's and 60's was in a transitionary period during which a two–sector economy consisting of a rural agricultural sector and an urban manufacturing sector was transformed into a one–sector industrial economy. The concomitant sharp increase in the number of households continuously stimulated domestic *aggregate* demand, and as a consequence *economy–wide* investment, too. Rapid economic growth in this period was basically sustained by domestic demand.

Investment was autonomous in relation to measures of aggregate economic activities such as real GNP. We also observed strong parallels in investment across various sectors during this period, which suggests that economy–wide rather than industry–specific factors spurred investment.

The situation changed dramatically around 1970. Here we stressed that internal factors were largely responsible for this change, but of course we realize the need to study more carefully the impacts of the oil shocks on the structural change.

In any event, the industrial structure became heavily tilted toward the machinery block, consisting of fablicated metals, general machinery, electrical machinery and transportation machinery. Simultaneously, in contrast to the period of rapid economic growth, the Japanese economy was becoming more dependent on exports in the 1970's and 80's. The ratios of exports (in volume terms) in 1984 to exports in 1970 are respectively 5.3, machinery broadly defined; 2.0 metals; 1.8 chemicals; 1.3 textiles; and 0.9 food. Naturally, the relative share of the machinery block in total manufacturing investment has increased over this period.

The points we discussed above are hard to verify by using standard econometric methods. We believe, however, that it is essential for us to keep these factors in mind before we proceed to undertake any econometric work. In the next section, we estimate industry investment functions.

2. Industry Investment Functions

In this section, we propose a theoretical model of investment and estimate these investment functions.

The Model

It is well known that the dynamic model of investment based on convex adjustment costs can be interpreted in the context of Tobin's q theory (1969): See, for example, Yoshikawa (1980) and Hayashi (1982). In terms of this theory, all the information relevant for investment decisions is conveniently summarized in a single variable q. Thus this approach is elegant in its simplicity, and it also easily lends itself to empirical implementation.

Despite these appealing features, several puzzles and problems associated with the q approach have arisen in empirical studies. (1) Without exception, empirical q equations contain lagged qs, or they show low Durbin–Watson statistics if no lags in q are introduced. Note that only current q should matter in the ordinary q theory. (2) The performance of empirical q equations is often reported to be 'poorer' than simple accelerator–type models: See Clark (1979). (3) In many countries investment is known to be highly correlated with current 'real' variables such as profits, profit rate or capacity utilization rate. Abel and Blanchard (1986), for example, include the profit rate together with q in their investment equations and find that current profit rate is always quite significant and that in some cases q even turns out to be insignificant. The Abel and Blanchard work is based on a carefully constructed *marginal* q series, and yet their findings are very similar to the results obtained relating investment to *average* q. Therefore they conclude it is very unlikely that the low explanatory power of q is due to the fact that average q is simply a poor proxy for the theoretically relevant marginal q: See also Gordon and Veitch (1986).

In an attempt to solve these problems associated with q theory, Ueda and Yoshikawa (1986) showed that the introduction of the two important factors — delivery lags for capital and differences in the time series properties of profit rates and discount rates — leads to a crucial modification of the standard q theory of investment. To be specific, under the assumptions made, it is shown that investment depends on distributed lags of profits rates and discount rates (or costs of capital) *separately*. In other words, we do not obtain a simple expression of q, which gives a particular constraint on two series of profits and costs of capital as the determinants of investment. Moreover investment is correlated more with profit rates than with discount rates if profit rates contain more permanent movements than discount rates. The reason is that with time to build, investment ordered today starts being productive only after a certain interval, and therefore rational investment decisions must regard temporary fluctuations in the discount rate as irrelevant.

The model presented below is similar to the one in Ueda and Yoshikawa (1986), and is based on the same observation that the distinction between permanent and transitory shocks in the determinants is important. In particular, here we explore the importance of such components of profits as sales, real wages and real energy costs in determining investment in Japan. Suppose that the firm maximizes the following discounted sum of future profits, V_t:

$$V_t = E[\sum_{j=0}^{\infty} \{\prod_{i=0}^{j} (1 + E(R_{t+1}|\Omega_{t+i-1}))^{-1} D_{t+j}|\Omega_{t-1}] \quad (1)$$

where the one–period discount rate R_t, and profits D_t are both taken as stochastic. $E(\cdot|\Omega_{t-1})$ stands for the expected value operator conditional on the information set as of time t–1. Ω_{t-1}.

We assume that profits D_t can be written as follows:

$$D_t = F(K_t, L_t) - w_t L_t - (1 + E(R_t|\Omega_{t-1}))^{-1}\phi(I_{t-k}). \quad (2)$$

$F(K_t, L_t)$ is the amount of output or sales, and K_t and L_t are capital stock and variable input, respectively. Production function F is assume to have the neoclassical properties. w_t is the real input price of L_t. As usual both w_t and L_t can be vectors. The last term in (2) is investment expenditures inclusive of adjustment costs. The presence of the discount factor simply indicates that payment on investment goods is made at the beginning of the period t. The function ϕ, reflecting increasing adjustment costs, is assumed to be convex:

$$\phi(0) = 0, \phi' > 0, \phi'' > 0. \quad (3)$$

An important feature of the model is the existence of a time–to–build or delivery lag (k periods). As a result, I_{t-k}, the order of investment goods for delivery in period t enters the ϕ function in period t. Here just for expository simplicity, it is assumed that investment expenditure is made when the project is completed and starts being productive, i.e. starts becoming a part of K_t. Noting the existence of a k period time to build, we have the following equation for capital accumulation:

$$K_{t+1} = K_t + I_{t-k}. \quad (4)$$

Next we assume a possible quantity constraint on the firm's sales.

$$Q_t \geqq F(K_t, L_t). \quad (5)$$

Q_t is exogenous stochastic sales. When the above inequality strictly holds, the firm is in the neoclassical regime whereas when the equality holds, the firm is in the quantity–constrained (or perhaps Keynesian) regime.

The firm maximizes (1) under constraints (2), (4), and (5) with respect to I_t and L_t. W_t, R_t, and Q_t, all stochastic, are taken as exogenous, together with suitable initial values of K_t. In general, the optimizing firm would anticipate a switch or even switches of two regimes in future. In the deterministic model, this possibility is explored by Precious (1986). In the present stochastic model, however, it is simply too difficult to analyze to obtain interesting economic implications. Therefore in what follows, we assume that the firm is in either the neoclassical or the quantity–constrained regime. we must analyze two possible regimes separately. As is usual, when the solution exists, it is unique under the strict concavity assumption.

Quantity–Constrained Regime

In this case, equality holds in constraint (5). The optimality condition for investment is given by

$$\phi'(I_t) = E(q_{t+k} \mid \Omega_{t-1}). \tag{6}$$

Here the variable q_{t+k} can be interpreted as Tobin's marginal q, and is as follows:

$$q_{t+1} = E[\sum_{i=0}^{\infty} \rho_{t+k}^{i+1} \frac{(w_{t+k+i})(\frac{\partial F}{\partial K_{t+k+i}})}{(\frac{\partial F}{\partial L_{t+k+i}})} \mid \Omega_{t+k-1}]. \tag{7}$$

For simplicity, we use the discount factor ρ_τ^{j+1} in (7):

$$\rho_\tau^{j+1} \equiv \prod_{i=0}^{j} \{1 + E(R_{\tau+i} \mid \Omega_{\tau+i-1})\}^{-1} \quad (j = 0, 1, 2, ..., \tau = t, t+1, ...). \tag{8}$$

We note that

$$\rho_t^1 \in \Omega_{t-1}, \rho_t^{j+1} \notin \Omega_{t-1} \qquad \text{for } j > 0$$
$$\rho_{t+i}^j \notin \Omega_{t-1} \qquad \text{for all } i, j > 0. \tag{9}$$

The marginal q (7) is the discounted sum of future savings in labor cost due to capital accumulation. In the present quantity–constrained regime, the firm can reduce labor cost and produce the same amount of output by increasing its capital stock. Note that q_{t+k} is stochastic as of time t. The optimality condition (6) then says that investment is determined in such a way that the expected value of q_{t+k} conditional on Ω_{t-1} is equal to the marginal adjustment cost of investment, $\phi'(I_t)$. Only in the degenerate case when k is zero, i.e., when there is no delivery lag or gestation period for investment, do we obtain the usual optimality condition involving the current nonstochastic q:

$$\phi'(I_t) = q_t. \tag{10}$$

With a lag to build, however, the q theory must be modified in the way indicated by (6). In particular, the current investment decision I_t, depends on the expected value of *future* marginal q evaluated at the beginning of time t. Alternatively, investment expenditure (or simply investment, in the usual sense of the term) at t, $INV_t \equiv \varphi(I_{t-k})$, depends on the *expected value* of q_t evaluated at the beginning of t–k, rather than on current q_t itself:

$$INV_t \equiv \varphi(I_{t-k}) = \psi[E(q_t \mid \Omega_{t-k-1})] \qquad \psi' > 0. \tag{11}$$

Our investment function is given by (7) and (11). To proceed further, it is useful to use the Cobb–Douglas illustration. Assume that the production function F is of the Cobb–Douglas type:

$$F(K, L) = K^{\alpha}L^{1-\alpha} \qquad (0 < \alpha < 1). \tag{12}$$

Then the marginal q, (7) is:

$$q_t = E[\sum_{i=0}^{\infty} (\prod_{j=0}^{i} \beta_{t+j}) W_{t+i} (\frac{Q}{K})_{t+i}^{\frac{1}{1-\alpha}} (\frac{1}{1-\alpha}) \mid \Omega_{t-1}] \tag{13}$$

where for notational simplicity, we define

$$\beta_s \equiv \rho'_s = \frac{1}{1 + E(R_s \mid \Omega_{s-1})}. \tag{14}$$

Although the right hand side of (13) is a nonlinear function of β_t, W_t, Q_t, and K_t, we know that its linearization is not a bad approximation: See, for example, Abel and Blanchard (1986) on this point. Choosing the stable solution, and substitution K_t out by (4), the linearized q_t becomes

$$\begin{aligned} q_t = \bar{q} + \sum_{i=0}^{\infty} [&a_i E(\beta_{t+i} - \bar{\beta} \mid \Omega_{t-1}) + b_i E(W_{t+i} - \bar{W} \mid \Omega_{t-1}) \\ &+ c_i E(Q_{t+1} - \bar{Q} \mid \Omega_{t-1})] \qquad (a_i, b_i, c_i > 0 \text{ for all } i). \end{aligned} \tag{15}$$

From (11), therefore, investment as of time t, INV_t, depends on

$$E(q_t \mid \Omega_{t-k-1}) = \bar{q} + \sum_{i=0}^{\infty} [a_i E(\beta_{t+i} - \bar{\beta} \mid \Omega_{t-k-1}) + b_i E(W_{t+i} - \bar{W} \mid \Omega_{t-k-1}) + c_i E(Q_{t+i} - \bar{Q} \mid \Omega_{t-k-1})]. \quad (16)$$

The most important point to note is that investment is affected by only those parts of β_t, W_t and Q_t that were *anticipated* k periods before. In other words, if Ω_{t-k-1} contains current and past values of β_t, W_t, and Q_t, i.e., β_{t-k-i}, W_{t-k-i}, and Q_{t-k-i} $(i \geq 1)$, then it is *not*, β_{t-k-i}, W_{t-k-i}, and Q_{t-k-i} but rather only the persistent or "permanent" components of these variables that affect investment. In the quantity–constrained regime, an increase in the discount factor (or a decrease in the discount rate), an increase in the real price of variable input, and/or an increase in demand, if "permanent", all raise investment. The empirical investment functions we will estimate are based on these observations.

Neoclassical Regime

In this regime, the inequality constraint (5) is not binding and the firm is assumed to take only the real price of variable input W_t, and discount factor β_t as given.

The optimality condition (6) remains unchanged, and therefore we obtain the same investment function (11). However the marginal q in this case is

$$q_t = E[\sum_{j=0}^{\infty} \rho_t^{j+1} \frac{\partial F}{\partial K_{t+j}} \mid \Omega_{t-1}]. \quad (17)$$

Here again we will use the Cobb–Douglas illustration. In the case of the Cobb–Douglas production function, after the linearization, (17) becomes

$$q_t = \bar{q} + \sum_{i=0}^{\infty} [\alpha_i E(\beta_{t+i} - \bar{\beta} \mid \Omega_{t-1}] - \beta_i E(W_{t+i} - \bar{W} \mid \Omega_{t-1})] \quad (18)$$

$$(\alpha_i, \beta_i > 0 \text{ for all i}).$$

As explained previously (equation (11)), investment at time t, INV_t, depends on $E(q_t | \Omega_{t-k-1})$:

$$E(q_t \mid \Omega_{t-k-1}) = \bar{q} + \sum_{i=0}^{\infty} [\alpha_i E(\beta_{t+i} - \bar{\beta} \mid \Omega_{t-k-1}) - \beta_i E(W_{t+i} - \bar{W} \mid \Omega_{t-k-1})]. \quad (19)$$

We note that in the present neoclassical regime, a permanent increase in W_t *lower* investment. This result is in sharp contrast to that obtained in the quantity–constrained regime. In the quantity–constrained regime, the firm is on a particular isoquant. Therefore an increase in W_t makes the firm substitute capital for labor and as a consequence raises investment. In the present neoclassical regime, on the other hand, the firm is on a particular factor price frontier. In this case, an increase in W_t lowers profits and thereby discourages investment. *Changes in the real price of variable input* L_t *have just the opposite effect on investment depending on whether the firm is in the quantity–constrained regime or in the neoclassical regime.*

Decomposition of Permanent and Transitory Shocks

Our theoretical model indicates the importance of the distinction between permanent and transitory shocks in Q_t, W_t, and β_t. Suppose that the information set as of t–k–1, Ω_{t-k-1}, contains Q_{t-k-i}, W_{t-k-i} and β_{t-k-i} $(i \geqq 1)$. Irrespective of whether the firm is in the quantity–constrained or the neoclassical regime, it is only the permanent components of Q_{t-k-i}, W_{t-k-i}, and β_{t-k-i} that affect investment. It is therefore necessary to decompose Q_{t-k-i}, W_{t-k-i}, and β_{t-k-i} into "permanent" and "transitory" components to obtain the investment function.

In current practice, there is no generally accepted method of doing such decompositions. The approach we use in this paper is the Beveridge–Nelson decomposition. For details, please refer to Beveridge and Nelson (1981); below we briefly summarize their technique.

Denote the observations on a nonstationary series by Z_t and its first difference by W_t so that $W_t = Z_t - Z_{t-1}$. If the W's are stationary in the sense of fluctuating around a fixed mean with stable autocovariance structure, then the Wold decomposition theorem implies that W_t may be expressed as

$$W_t = \mu + \epsilon_t + \lambda_1 \epsilon_{t-1} + \dots \tag{20}$$

where μ is the long–run mean of the W series, the λ_i are constants, and the ϵ's are white noises with mean zero and variance σ^2.

Then we consider the k–period ahead forecast of Z_t, $E(Z_{t+k} \mid \Omega_t)$ in the case where Ω_t

contains only Z_{t-i} ($i \geqq 0$). It can be shown that when k is large, $E(Z_{t+k} \mid \Omega_t)$ is approximately as follows:

$$E(Z_{t+k} \mid \Omega_t) \cong k\mu + Z_t + (\Sigma_1^{\infty}\lambda_1)\epsilon_t + (\Sigma_2^{\infty}\lambda_i)\epsilon_{t-1} + \ldots \tag{21}$$

It is apparent from (21) that the forecast profile is a linear function of forecast horizon k with slope equal to μ, the rate of drift of the series, and a "level" (algebraically the intercept) which itself is a stochastic process. Beveridge and Nelson suggest that it is natural to interpret this level as the *permanent* component of Z_t. Denoting this level by Z_t^*, we have

$$Z_t^* = Z_t + (\Sigma_1^{\infty}\lambda_i)\epsilon_t + (\Sigma_2^{\infty}\lambda_i)\epsilon_{t-1} + \ldots \tag{22}$$

It can be shown that this permanent component of Z_t, Z_t^* is the following random walk with a drift.

$$Z_t^* = Z_{t-1}^* + \mu + (\Sigma_0^{\infty}\lambda_i)\epsilon_t, \qquad \lambda_0 \equiv 1. \tag{23}$$

The *transitory* part of Z_t, $\tilde{Z}_t$ is by definition

$$\tilde{Z}_t \equiv Z_t - Z_t^* = -(\Sigma_1^{\infty}\lambda_i)\epsilon_t - (\Sigma_2^{\infty}\lambda_i)\epsilon_{t-1} - \ldots \tag{24}$$

The permanent part, Z_t^*, is a random walk and therefore certainly consists of permanent shocks, but how transitory the $\tilde{Z}_t$s are depends on the moving average coefficients, λ_i, of $Z_t - Z_{t-1}$. In our theory, those parts of β_{t-k-i}, W_{t-k-i}, and Q_{t-k-i} ($i \geqq 1$) which do not affect investment at time t (INV_t), consist of shocks which disappear within k periods, i.e. delivery lags or time to build. Therefore the Beveridge–Nelson decomposition does not conceptually exactly correspond to the decomposition of our theory. Having noted this reservation, however, we will apply their method in what follows.

Estimated Equations

Our investment function is equation (11). The expected marginal q in (11) is given by (16) in the case of the quantity–constrained regime and by (19) in the case of the neoclassical regime.

We assume that the firm's information set Ω_{t-k-1} contains β_{t-k-i}, W_{t-k-i}, and Q_{t-k-i} ($i \geqq 1$) and also that in its investment decisions the firm responds only to what it perceives as the permanent components of the relevant variables. There is no knowing how the firm actually extracts the permanent components from the observed series. We simply apply the Beveridge–Nelson decomposition to the explanatory variables.

We estimate investment functions for Japanese manufacturing industries using quarterly data (1971.I–1986.II). Since investment in the 70's and 80's seems to be governed by micro–specific factors, as we argued in Section 1, it makes sense to estimate industry investment equations rather than an aggregates equation. Investment figures, I_t, are from the "newly acquired fixed assets" series compiled by the Ministry of Finance. For demand variable Q_t, we use an index of shipments, S_t, published by the Ministry of Trade and Industry. We identify labor and energy as variable inputs and define the real wage, RW_t, and the real energy price, RE_t, as follows:

$$RW \equiv \frac{(\text{Index of Total Payment in Cash to Regular Employees})}{(\text{Index of Total Working Hours of Regular Employees}) \cdot (\text{WPI of Demestic Manufactured Products})} \tag{25}$$

$$RE \equiv \frac{(\text{Index of Prices of Imported Oil, Coal and Natural Gas})}{(\text{WPI of Domestic Manufactured Products})}. \tag{26}$$

We also use the price of investment goods, PI_t, which is defined as

$$PI \equiv \frac{(\text{WPI of Investment Goods})}{(\text{WPI of Domestic Manufactured Products})}. \tag{27}$$

Note that PI is implicit in function ψ in (11), but that investment is a decreasing function of PI. For simplicity, we assume that RE and PI are the same across industries.

We tried several cost of capital or interest rate variables, corresponding to β, in our preliminary set of estimations. However the results were not satisfactory, and therefore we dropped them from the final set of estimations presented below.

As the first step, we applied the Beveridge–Nelson decompositions to S_t, W_t, RE_t, and PI_t. For this purpose, we first fitted the ARIMA (Autoregressive Integrated Moving Average) models to S, RW, RE, and PI. All the variables including I are in *natural logarithms*. We use the same notation for the logs of the variables. The final models were selected by Akaike's Information Criterion (AIC). It turned out that all the variables are first–order integrated.

Table 3: Estimated Industry Investment Equations

Nonferrous Metals

$$I_t = \underset{(-0.79)}{-1.36} + \underset{(3.98)}{1.54}\, S^*_{t-4} - \underset{(-3.63)}{1.81}\, \tilde{S}_{t-4} + \underset{(1.90)}{0.18}\, RE^*_{t-4} + \epsilon_t$$

$$\hat{\rho} = 0.55, \qquad R^2 = 0.75$$

Chemicals

$$I_t = \underset{(3.36)}{3.55} + \underset{(3.23)}{0.82}\, S^*_{t-4} - \underset{(-1.40)}{0.85}\, \tilde{S}_{t-4} + \underset{(2.25)}{0.15}\, RE^*_{t-1} + \epsilon_t$$

$$\hat{\rho} = 0.79, \qquad R^2 = 0.89$$

General Machinery

$$I_t = \underset{(-5.42)}{-4.37} + \underset{(2.93)}{0.58}\, RW^*_{t-3} + \underset{(9.06)}{1.83}\, S^*_{t-3} + \underset{(0.39)}{0.22}\, \tilde{S}_{t-2} + \underset{(2.26)}{0.16}\, RE^*_{t-1} + \epsilon_t$$

$$\hat{\rho} = 0.19, \qquad R^2 = 0.91$$

Electrical Machinery

$$I_t = \underset{(-1.98)}{-1.52} + \underset{(2.35)}{0.61}\, RW^*_{t-3} - \underset{(-2.26)}{2.70}\, \tilde{RW}_{t-3} + \underset{(11.0)}{1.48}\, S^*_{t-3} - \underset{(-1.43)}{0.64}\, \tilde{S}_{t-3} + \epsilon_t$$

$$\hat{\rho} = 0.62, \qquad R^2 = 0.98$$

Transportation Equipment (Automobile)

$$I_t = \underset{(1.50)}{12.1} + \underset{(2.80)}{1.00}\, RW^*_{t-5} + \underset{(5.25)}{1.67}\, S^*_{t-5} - \underset{(-2.72)}{0.47}\, RE^*_{t-5} + \underset{(3.05)}{0.90}\, \tilde{RE}_{t-5} - \underset{(-1.82)}{2.99}\, PI^*_{t-5} + \epsilon_t$$

$$\hat{\rho} = 0.56, \qquad R^2 = 0.91$$

Pulp, Paper & Paper Products

$$I_t = \underset{(0.54)}{0.87} + \underset{(2.57)}{0.79}\, RW^*_{t-4} + \underset{(1.13)}{0.42}\, S^*_{t-1} - \underset{(-0.99)}{0.84}\, \tilde{S}_{t-4} + \underset{(1.20)}{0.11}\, RE^*_{t-1} + \epsilon_t$$

$$\hat{\rho} = 0.69, \qquad R^2 = 0.77$$

Based on these models, we decompose S_t, RW_t, RE_t, and PI_t into the Beveridge–Nelson permanent components denoted by S_t^*, RW_t^*, RE_t^*, PI_t^*, and the transitory components, $\tilde{S}_t$, $\tilde{RW}_t$, $\tilde{RE}_t$, $\tilde{PI}_t$.

In the second step, we estimated industry investment equations using S_t^*, $\tilde{S}_t$, RW_t^*, $\tilde{RW}_t$ and so on rather than S_t, RW_t, RE_t, PI_t as explanatory variables. In addition to the six industries we have here, the steel and textile industries were included in our preliminary set of estimations. However, we did not succeed in obtaining satisfactory results for these two industries. It is very interesting that these two industries were proto–typical declining industries over the sample period. For example, the Japanese steel industry's capacity was a hundred million tons in 1971. Thereafter the industry never tried to expand its capacity and has consistently suffered from overcapacity. For the purpose of increasing capacity, net investment was always negative over the sample period. Due to the appreciation of the yen and stiffer competition from the NIC's in recent years, this process has accelerated. Therefore it may not be so surprising after all that our theory of investment is not applicable to the Japanese steel industry. The situation is very similar in the textile industry.

The final equations estimated for the six industries are shown in Table 3. In every case, we tried lags of from one to six periods. These lags correspond to delivery lags, k, and they will differ for structures and machinery. The existing empirical studies suggest a lag of from a year to a year and a half. The method of estimation is generalized least squares; to be specific, full transform estimation (Cf. Harvey (1981, Ch. 6)) was employed.

Except for the pulp and paper industry, permanent shipments, S^*, has significant effects on investment. The elasticity is 0.8 for the chemical industry, but for other industries it is about 1.5. In theory, transitory shipments, $\tilde{S}$, should not affect investment. On the whole, only S_t^* is significant. The exception is the nonferrous metal industry. We have already noted, however, that $\tilde{S}$ based on the Beveridge–Nelson decomposition does not exactly correspond to its counterpart in theory.

As for real factor prices, their effects on investment are mostly positive. This is consistent with the above result that shipments significantly affects investment. These results suggest that firms were in the quantity–constrained regime over the sample period. An increase in real wages raises investment in the three machinery industries and the pulp and paper industry with the elasticities ranging from 0.6 to 1.0. An increase in real energy costs, on the other hand, raises investment in such industries as nonferrous metals, chemicals, general machinery and transportation machinery. The elasticities in this case are smaller than the wage elasticities and are about 0.2.

3. Concluding Remarks

The Japanese economy experienced a major structural change around 1970. Accordingly the behavior of fixed investment also changed at the same time. In contrast to the rapid economic growth period in the 1950's and 60's when investments in various industries were simultaneously spurred by aggregate domestic demand, investment in the 1970's and 80's seemed to be generated by industry–specific factors. This suggests that analysis of investment should be undertaken on an industry level basis for the 1970's and 80's. To be more specific, the machinery industries, which underwent dramatic IC–related technological innovations, became more and more important in the 70's and 80's. Since the products of these industries were geared more toward foreign markets, the Japanese economy as a whole became more dependent on exports over this period.

As the reason for the structural change around 1970, we emphasized in Section 1 the importance of domestic factors such as population flows, but we realize the need to study more carefully the impact of the oil shocks on this phenomenon.

In Section 2, we proposed a theory of investment similar to Ueda and Yoshikawa (1986). We showed that the usual q theory must be modified substantially when delivery lags exist, and also that in this case distinguishing between permanent and transitory shocks becomes essential. Moreover we showed that one obtains a very different investment function depending on whether the firm is in the quantity–constrained regime or in the neoclassical regime. Using quarterly (1971.I – 86.II) data on Japanese manufacturing industries, we estimated industry investment functions. Although we did not succeed in obtaining good results for two prototypical declining industries, steel and textile, we did get reasonable results for six industries including three in the machinery sector, the sector which led the Japanese economy over the sample period.

The estimation results are on the whole in accord with our theory. In particular, the significant effects of shipments and positive effects of real factor prices imply that firms were in the quantity–constrained regime. Shapiro (1986) argued that if the production function is subject to technological shocks, then output and profits and hence also investment all simultaneously increase. In this case, shipments may appear to drive investment even if the firm is in the neoclassical regime. We argue, however, that (1) the important technological improvements are embodied in new investment and (2) the short–run fluctuations of labor productivity are demand–induced given the firm's labor hoarding; therefore Shapiro's argument is extremely unconvincing. In any event, if firms are demand–constrained, the straightforward application of the neoclassical factor price frontier is questionable. In the Japanese economy, our results indicate that increases in real wages or real energy costs raised rather than lowered investment.

Whether the economy is in the quantity–constrained or the neoclassical regime is also an important issue in business cycle theory. Recent real business cycle theory of Kydland and Prescott (1982), Long and Plosser (1983) and others take the view that the economy is always in the neoclassical equilibrium. In the context of the Japanese economy, Yoshikawa and Ohtake (1987) presented some evidence that real shocks are important but they are mostly demand shocks rather than the supply side technological shocks whose role is emphasized by real business cycle theorists. The estimation results of investment functions in this paper also suggest that the economy was demand–constrained over the period.

Finally closely related to the above problem is the issue of how important are permanent shocks vis–a–vis aggregate fluctuations. Statistically this problem has been analyzed under the heading of tests for the existence of unit roots in economic time series: See Nelson and Plosser (1982), Harvey (1985) and Rose (1986). The theory of investment advanced in Section 2 has the significant implications for this issue. No doubt the economy is subject to both permanent and transitory shocks. It may be the case that economic agents processing two different types of shocks respond only to permanent shocks, and as a consequence most economic time series consist largely of permanent shocks. It seems reasonable to believe that real shocks are normally more persistent than monetary shocks, and therefore real shocks play the dominant role in aggregate fluctuations (Cf. Ueda and Yoshikawa (1986) on this point). As we discussed above, however, whether these real shocks are demand shocks or supply shocks is an entirely separate issue which awaits further investigation.

* I acknowledge with gratitude the excellent research assistance of Yoshiyuki Takeuchi. I am also grateful to David W. Campbell for his editorial assistance, and to H. Yamawaki and other participants of the Conference for their helpful comments. The research reported here was supported by the Kikawada Foundation.

REFERENCES

Abel, B. A., and O. J. Blanchard: "The Present Value of Profits and Cyclical Movements in Investment," *Econometrica*, 54 (1986), 249–273.

Beveridge, S., and C. Nelson: "A New Approach to Decomposition of Economic Time Series into Permanent and Transitory Components with Paticular Attention to Measurement of the Business Cycles," *Journal of Monetary Economics*, 7 (1981), 151–174.

Bruno, M., and J. Sachs: *Economics of Worldwide Stagflation*. Cambridge, Mass.: Harvard University Press, 1985.

Clark, P.: "Investment in the 1970's," *Brookings Papers on Economic Activity*, (1979), 73–124.

Gordon, R. J., and J. M. Veitch: "Fixed Investment in the American Business Cycle, 1919–83," in *The American Business Cycle*, ed. by R. J. Gordon. Chicago: Unversity of Chicago Press, 1986.

Harvey, A. C.: *Econometric Analysis of Time Series*. Oxford: Philip Allan, 1981.

———: "Trends and Cycles in Macroeconomic Time Series," *Journal of Business & Economic Statistics*, 3 (1985), 216–227.

Hayashi, F.: "Tobin's Marginal q and Average q: A Neoclassical Interpretation," *Econometrica*, 50 (1982), 213–224.

Horie, Y., S. Naniwa, and S. Ishihara: "The Changing Pattern of Business Fluctuations," (in Japanese) *Kinyu Kenkyu*, 6 (1987), 13–60.

Kuroda, M., K. Yoshioka, and M. Shimizu: "Economic Growth: Factor Analysis and Interindustry Relations," (in Japanese) in *A Macroeconomic Analysis of the Japanese Economy*, ed. by K. Hamada et. al. Tokyo: Tokyo University Press, 1987.

Kydland, F. E., and E. C. Prescott: "Time to Build and Aggregate Fluctuations," *Econometrica*, 50 (1982), 1345–1370.

Long, J., and C. Plosser: "Real Business Cycles," *The Journal of Political Economy*, 91 (1983), 39–69.

Nelson, C., and C. Plosser: "Trends and Randam Walks in Macroeconomic Time–series," *Journal of Monetary Economics*, 10 (1982), 139–162.

Norsworthy, J. R., and D. H. Malmquist: "Input Measurement and Productivity Growth in

Japanese and U.S. Manufacturing," *American Economic Review*, 73 (1983), 947–967.

Precious, M.: "Demand Constraints, Rational Expectations and Investment Theory," *Oxford Economic Papers*, 37 (1985), 576–605.

Rose, A.: "Four Paradoxes in GNP," *Economics Letters*, 20 (1986), 137–141.

Shapiro, M.: "Investment, Output and the Cost of Capital," *Brookings Papers on Economic Activity*, (1986), 111–164.

Tobin, J.: "A General Equilibrium Approach to Monetary Theory," *Journal of Money, Credit and Banking*, 1 (1969), 15–29.

Ueda, K., and H. Yoshikawa: "Financial Volatility and the q Theory of Investment," *Economica*, 53 (1986), 11–27.

Yoshikawa, H.: "On the q Theory of Investment," *American Economic Review*, 70 (1980), 739–743.

———, and F. Ohtake: "Postwar Business Cycles in Japan: A Quest for the Right Explanation," *Journal of Japanese and International Economies*, 1 (1987), 373–407.

Testing the Sensitivity of Q Investment Equations to Measurement of the Capital Stock

Stephen Bond and Michael Devereux[1]

Institute for Fiscal Studies, London

Introduction

The Q model of investment has recently become a common vehicle for the empirical investigation of the investment decisions of firms. There are two main reasons for its popularity. The model can be derived from an explicit optimisation framework, and it rationalises why expectations play a crucial role. In addition, under conditions of perfect competition and constant returns to scale, unobservable expectations regarding profits and dividends are conveniently summarised by the observable market value of capital (Hayashi (1982)).

Despite these apparent advantages, Q models have had limited empirical success. In particular, Q tends to have low explanatory power and either serial correlation or dynamic structures involving the lagged dependent variable are common. In addition, the coefficient on the Q term is generally found to be very low which, according to the theory underlying the model, suggests that adjustment costs are extremely high.

There are several possible reasons for these findings. One which has recently been examined is that the results stem from problems of aggregation. Testing the theory on individual company data over a number of years offers several advantages over time-series estimation. The large time-series and cross-section variation available in company data should improve the accuracy of parameter estimates (cross-section variation is less useful in normal formulations of the cost of capital, which does not vary across companies). In addition, some important variables, notably the market value of the company, can be accurately measured. However, two recent papers (Blundell, Bond, Devereux and Schiantarelli (1987) and Hayashi and Inoue

(1987)) have found that although Q is a significant determinant of investment at the company level, most of the empirical problems with aggregate Q models are also present using micro data.

Another reason for this lack of empirical success may be due to measurement error in Q. Since Q is essentially the ratio of the market value of capital to the replacement value of the capital stock, these two variables give particular cause for concern. The difficulty concerning the market value of capital is that, although the observed market value may be measured precisely, it may be affected by financial volatility in the stock market. Thus, there may be substantial movements in the share price of a company which either exaggerate, or are completely independent of, changes in its underlying economic prospects. If this is so, such speculative bubbles or fads would bias downwards the coefficient on Q. However, this argument received little empirical support from Hayashi and Inoue (1987).

In this paper we therefore examine the possibility that the empirical problems with Q investment equations are due to problems concerning the measurement of the replacement value of the capital stock. The literature on measurement of the capital stock has mostly been concerned with questions such as the average life of assets and the appropriate depreciation rate to use in empirical studies (see, for example, Hulten and Wykopf (1981), Griffin (1976), King and Fullerton (1984) and Wadhwani and Wall (1986)). However, one particular period which causes concern is 1974/5 - the period of the first oil price shock. It has been suggested (for example, by Baily (1981) and Summers (1981)) that the quadrupling of the oil price in this period reduced the value of the existing capital stock, which was too heavily dependent on oil. For example, Summers argued that "the increase in energy prices ... that occurred almost certainly has hastened the obsolescence of some capital goods. Thus depreciation has been understated in national accounts". By implication, the value of the capital stock was overestimated.

Aggregate studies have little choice in which measure of the capital stock to use - an estimate is presented in national accounts which provides the basis of most studies. However, in this paper we exploit the wider range of data available in company accounts. This accounting data includes the historic cost value of the net and gross capital stock of the company (unadjusted for inflation) in each year, supplemented in the UK in the early 1980s by accountants' estimates of the replacement value. In addition, some information on scrapping is available.

In section 2 we present three methods of estimating the replacement cost of the capital stock over a reasonably long time period, dating from the early 1970s. First, we estimate its value using the standard perpetual inventory method, using historic cost data and published estimates of depreciation rates and price movements. We then compare this measure with one derived from a similar approach, but which makes use of the replacement cost data from company accounts and is less susceptible to measurement problems associated with the energy price rises in the mid-1970s. Finally, we use a method proposed by Mairesse and Dormont (1985) which is closely related to historic cost accounting data and does not require prior knowledge of the depreciation rate. In section 3 we compare the performance of these three measures in a Q investment equation. Section 1 briefly outlines the framework of the Q model and section 4 concludes.

1. The basic Q model

A detailed derivation of the basic Q model is available in several other places (see, for example, Hayashi (1982), Poterba and Summers (1983) and Blundell, Bond, Devereux and Schiantarelli (1987)). In this paper we therefore present a more heuristic summary, paying particular attention to the role of the capital stock.

The standard Q model states that, under perfect competition, constant returns to scale and efficient financial markets, and in the absence of taxation, the optimal ratio of investment to the capital stock has a one-to-one relationship with "Q", defined as:

$$Q_t = \frac{p_t^I}{p_t}(q_t - 1) \tag{1}$$

where p^I_t is the price of investment goods in period t, p_t is the price of output in period t, and q_t is the ratio of the stock market valuation of capital in period t to the replacement cost value in period t of the capital stock held at the end of period t−1. The numerator of q_t has several components. In particular q_t is defined as:

$$q_t = \frac{V_t + H_t - Z_t}{(1-\delta)p_t^I K_{t-1}} \tag{2}$$

where V_t is the market value of the firm, Z_t is the value of assets other than capital held by the firm, H_t is the present value of all future cash flows associated with debt (including interest payments and additional funds derived from new debt), all in period t, δ is the rate of economic depreciation and K_{t-1} is the physical capital stock at the end of period $t-1$. It is assumed that the other assets (such as inventories) face no adjustment costs. They are deducted from the total market value of the firm in order to arrive at the market's valuation of the capital stock.

Three adjustments to the basic definitions in equations (1) and (2) must be made in order to introduce company taxation into the model[2]. First, the "effective" output price becomes $(1-\tau_t)p_t$, where τ_t is the "adjusted" corporation tax rate in period t. This "adjusted" corporation tax rate is not necessarily equal to the statutory tax rate. This is due to asymmetries within the UK corporation tax system, whereby a company with negative taxable profits cannot claim an immediate rebate, but must carry forward the loss to offset against subsequent profits. In such cases the "adjusted" rate of tax is therefore reduced by discounting until the firm resumes a tax paying position. Second, the "effective" price of capital goods becomes $(1-n_t)p^I{}_t$, where n_t is the present value of all tax depreciation allowances which arise as the result of a unit increase in investment. Again, this present value depends on whether, and in which subsequent periods, the firm is in a position of tax exhaustion, i.e. whether it must carry forward losses to subsequent periods. This is because the benefit of an allowance cannot be claimed until a later period and so must be discounted more heavily. Third, the market value, V_t, reflects the present value of future depreciation allowances that can be claimed on past investment (call this A_t). This sum must therefore be deducted from the numerator of q_t in (2) to find the market valuation of the capital stock.[3]

Putting these factors together, we can therefore summarise the Q model as:

$$\left(\frac{I}{K}\right)_t = \alpha + \beta Q_t + \epsilon_t \tag{3}$$

where

$$Q_t = \frac{(1-n_t)p_t^I}{(1-\tau_t)p_t}\left\{\frac{V_t + H_t - Z_t - A_t}{(1-\delta)(1-n_t)p_t^I K_{t-1}} - 1\right\} \quad . \tag{4}$$

The interpretation of the constant, α, is that rate of investment $(I/K)_t$ at which adjustment costs are zero. This is sometimes assumed to be the level of replacement investment. Strictly, β and the stochastic shock, ϵ_t, are also parameters of the adjustment cost function, with β being inversely related to the size of adjustment costs.

For the purposes of the analysis of this paper, the role played by the capital stock in the model (3) and (4) should be noted. The Q model uses a common expression for the equation of motion of the capital stock, namely

$$K_t = (1-\delta)K_{t-1} + I_t \tag{5}$$

This equation highlights the crucial role of the depreciation rate, δ. The three methods of estimating the replacement cost value of the capital stock set out below each use different values of δ (although the estimates of δ used in two of the methods only differ in 1974/5 when we consider there was unusually high obsolescence due to energy price movements). The methods also differ in their starting point for the perpetual inventory method (in two cases) and in their adjustments for differential price increases. The three methods therefore provide different estimates of K, which in turn provides alternative estimates of both Q and the rate of investment (I/K) in each period.

2. The data and some summary statistics

The principal data requirements for the estimation of the model developed above are cross-section and time series data for tax adjusted marginal Q and the gross investment rate. Two sources for this data have been used: company accounting data, from Datastream International, and share price and related data from the London Share Price Database.

Datastream provides company accounting data for all UK quoted companies from 1968. Currently we have access to periods up to 1985. The London Share Price Database offers a similar coverage. We selected from Datastream a sample of industrial and commercial companies, whose main activity is in the manufacturing sector, and which also had data available from the London Share Price Database and industry level price data. Excluding companies with less than twelve complete records, we were left with an unbalanced sample of companies with the following number of records:

No of companies	No of records per company
108	14
65	13
18	12

The detailed computation of the variables used is described in the data appendix. The principal complexities involve estimating the replacement cost values of the capital stock and the tax parameters. Since we do not have data on the physical size of new investment, I, nor the physical size of the capital stock, K, we estimate their ratio, the dependent variable, from estimates of their respective replacement cost values, $p^I I$ and $p^I K$. Since $p^I I$ is available from data, we require an estimate of $p^I K$, the replacement value of the capital stock. A similar term is required for the denominator of Q.

We calculate three estimates of the replacement value of the capital stock, making use of different information available in the data. The first is the commonly used perpetual inventory method. This takes an initial starting value and thereafter calculates the value for each succeeding year by depreciating the value in the previous year, augmenting it in line with specific price changes and adding new investment. One difficulty with this method is the lack of a split of recorded investment data by type of asset. The approach used is to split total investment between each asset using data on the change in the gross historic cost value of that asset. The original value of the replacement value, for 1968, was taken to be equal to the historic cost at that date. To reduce measurement error as a result of this assumption we use only data from 1971 in estimation. We refer to this approach as method A.

Method A assumes that the rate of economic depreciation, δ, is constant over time. It therefore makes no adjustment for unusually high obsolescence as a result of the 1974 energy price shocks. In addition, it tends to overestimate replacement cost values in the early 1980s relative to data reported in current cost company accounts. A second method, B, essentially assumes that the difference between the values derived from method A and the data in the early 1980s are due to increased obsolescence in 1974/5. It again uses the perpetual inventory method, but from a different starting point: the replacement cost estimates for the earliest year shown in the data (which is normally 1981). Estimates for earlier years are derived by applying the same approach as method A, but working backwards rather than forwards. These estimates were used back to 1975. Up to 1973 method A estimates are used. 1974 values were constructed as a weighted average (depending on the accounting year end) of the two estimates provided by methods A and B. Thus, the high obsolescence due to energy price shocks is reflected partly in the 1974 values, and fully by 1975. An advantage of method B is that, since it uses actual data from the early 1980s and does not depend on pre-1974 values, it is likely to produce better estimates of the replacement cost both for the early 1980s and late 1970s.

The perpetual inventory methodology does not use any of the information regarding depreciation in company accounts. Rather the depreciation rate is imposed (estimates in this paper are taken from King and Fullerton (1984)). However, company accounts do provide an estimate of the historic cost value of the net capital stock, from which depreciation has already been deducted. These figures need only an adjustment for price movements in order to estimate the replacement value. Following the approach used by Mairesse and Dormont (1985), our third method, C, makes such price adjustments by reference to the average age of the net capital stock. By comparing different estimates with current cost accounts data, Bond and Daniels (1988) have estimated this average age to be 3 years. To estimate replacement values, a simple adjustment to the historic cost values may be made, multiplying them by the increase in prices over the previous 3 years.

Figures 1 and 2 summarise the data constructed according to these algorithms. Figure 1 shows the sample mean replacement value of the capital stock at current prices in each year and for each method[4]. As would be expected, method A results in higher estimates because no account is taken of unusually high obsolescence in the mid-1970s. The divergence between method A and methods B and C grows over time. Figure 2 shows the sample mean of the investment rates calculated from each of the methods. Again as would be expected, method A results in lower estimates. However, all three measures tend to move together.

The tax parameters used in calculating Q vary from company to company for three reasons. First, the value of investment allowances in any period depends on the asset split of investment in that and previous periods. Second, because companies declare their accounting results at different points in the year, and statutory tax rates may change, then companies may face a different average tax rate. Third, we take into account the effects of tax exhaustion. We use estimates of the periods in which each firm is tax-exhausted taken from a model of the UK corporation tax system which applies tax rules to company accounting data to estimate tax liabilities (see Devereux (1986)). With carry forward provisions in the tax system, tax liabilities are simply delayed until the firm once more becomes a taxpayer. Our approach is therefore to assume that all the tax effects in the model are simply postponed until the firm ends its period of tax exhaustion and they are consequently discounted over this additional period of time.

Figure 1. Alternative measures of Capital Stock replacement cost

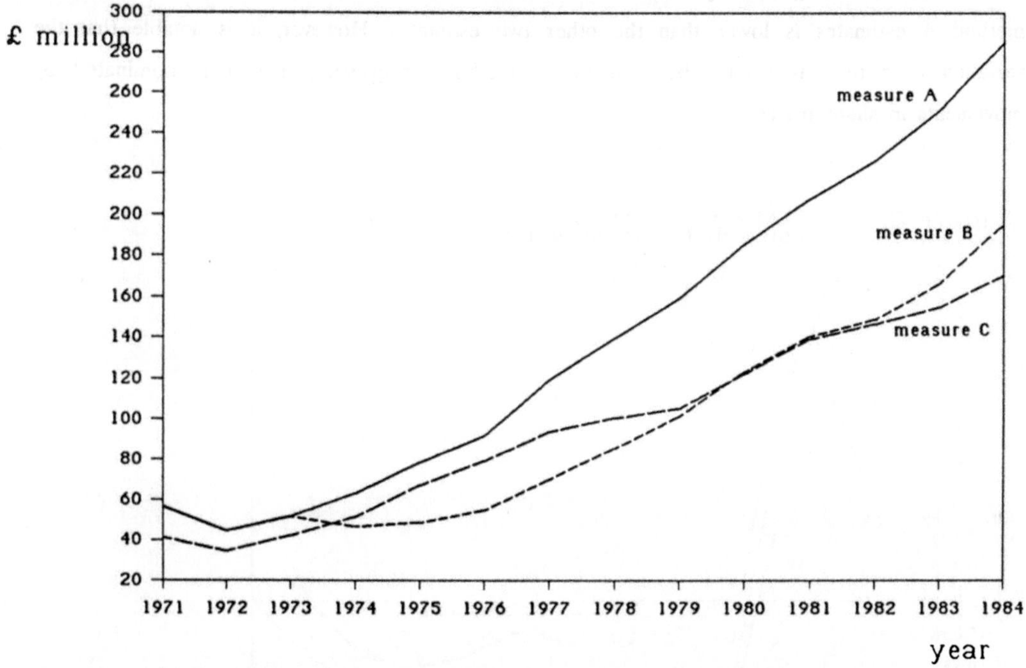

Figure 2. Alternative Measures of Investment Rate using different measures of K

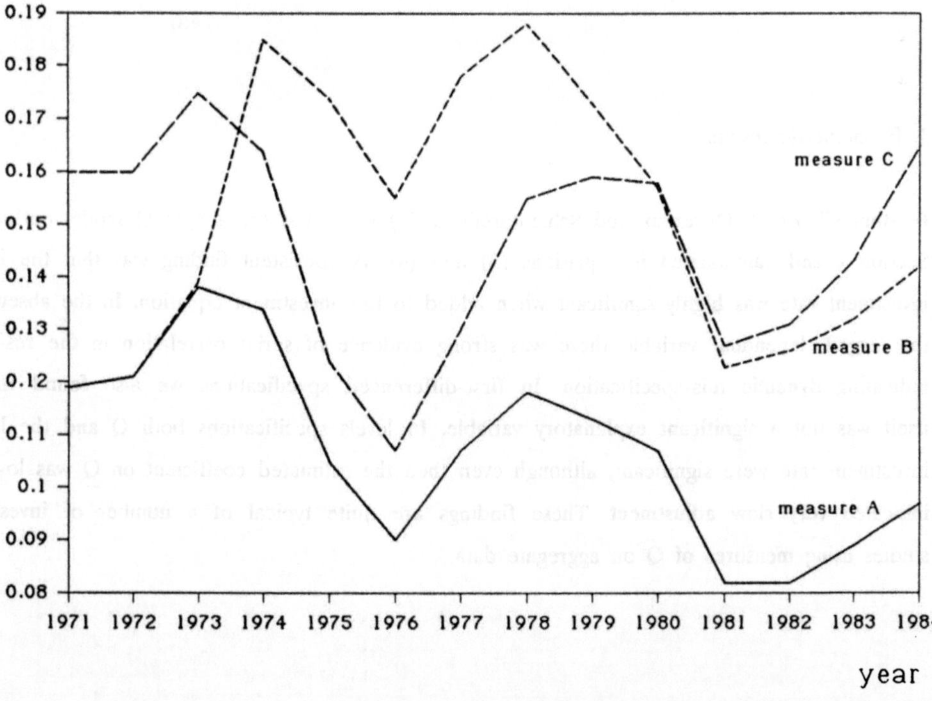

Figure 3 shows the sample mean of tax adjusted marginal Q for each year and each of the 3 methods of calculating the replacement value of the capital stock. The value of Q using the method A estimates is lower than the other two estimates. However, it is notable that the variation over time is similar between the 3 methods, suggesting that it is dominated by movements in share prices.

Figure 3. Alternative Measures of Q using different measures of K

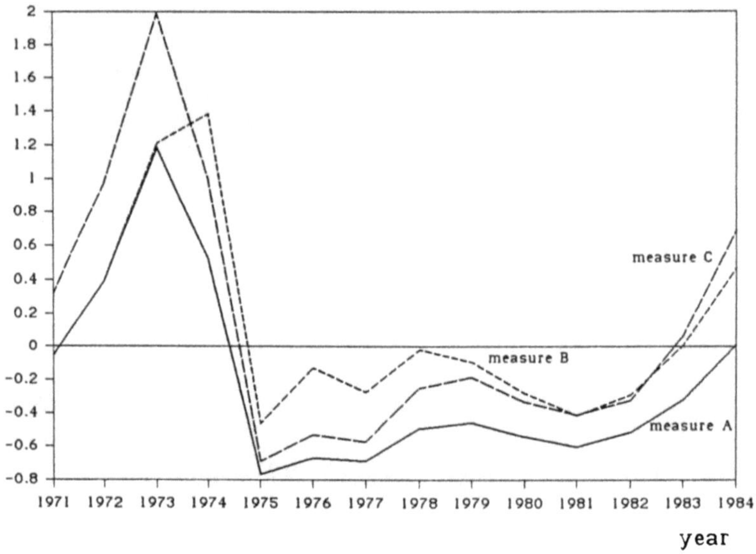

3. Econometric results

In Blundell, Bond, Devereux and Schiantarelli (1987) we tested the simple Q model outlined in Section 1 and summarised in equations (3) and (4). A consistent finding was that the lagged investment rate was highly significant when added to this investment equation. In the absence of the lagged dependent variable there was strong evidence of serial correlation in the residuals, indicating dynamic mis-specification. In first-differenced specifications we also found that Q itself was not a significant explanatory variable. In levels specifications both Q and the lagged investment rate were significant, although even then the estimated coefficient on Q was low and indicated very slow adjustment. These findings are quite typical of a number of investment studies using measures of Q on aggregate data.

One hypothesis that could explain these results is that Q is measured with error. We have noted that the replacement value of net fixed capital is notoriously difficult to measure. Bond and Daniels (1988) found that different procedures commonly used could result in widely divergent estimates, and these estimates could differ from those reported in current cost company accounts by factors of 2 and more. It is important to know whether the Q investment model is sensitive to the particular capital measure used, and whether some measures appear to perform better than others in investment equations, in the sense that the empirical problems noted with the Q approach are avoided or reduced.

In the previous section we described three approaches to measuring net fixed capital for a sample of UK companies. In their investment study Blundell, Bond, Devereux and Schiantarelli (1987) employed the standard perpetual inventory method (measure A) in constructing their measure of Q. The simplest specification that they found to be data conformable took the form:

$$\left(\frac{I}{K}\right)_{it} = \alpha_i + \alpha_t + \beta Q_{it} + \gamma\left(\frac{I}{K}\right)_{i,t-1} + \epsilon_{it} \qquad (6)$$

where i denotes companies and t denotes years. In panel A of Table 1 we report several estimates of this equation, where the data have again been constructed using measure A of the capital stock. Our sample of 191 companies is slightly smaller than that used in Blundell, Bond, Devereux and Schiantarelli (1987), the reason being that we were unable to construct capital stock measures B and C for all the companies included in the earlier study.

The presence of unobservable, firm-specific influences (α_i) on investment rates complicates the estimation of models like (6) in samples where the number of companies is large and the number of time periods is small. Taking first differences of the equation removes the firm effects but introduces a moving average error (ϵ_{it} - $\epsilon_{i,t-1}$) which necessitates the use of instrumental variables. The lagged dependent variable is clearly correlated with the lagged error ($\epsilon_{i,t-1}$). Even if Q_{it} is predetermined (cov(Q_{it}, ϵ_{it}) = 0, which is our maintained assumption since both the market value and replacement value of capital are measured prior to period t), it is likely to be correlated with past errors and so ΔQ_{it} should also be instrumented in the differenced equation. Measurement error problems may also be exacerbated by differencing. In column (i) we report an instrumental variables estimator of the equation in first differences,

TABLE 1 Q INVESTMENT EQUATIONS, 1975–84

A) Dependent variable is $(I/K^A)_{it}$ Standard deviation = 0.058

	(i) IV Differences	(ii) Within Groups	(iii) OLS Levels	(iv) Within Groups	(v) OLS Levels	(vi) GLS
$Q^A{}_{it}$	0.0049	0.0160	0.0116	0.0142	0.0153	0.0147
	(0.70)	(5.06)	(5.88)	(4.76)	(7.32)	(5.74)
$(I/K^A)_{i,t-1}$	0.2183	0.1203	0.4366	0.22*	0.22*	0.2173
	(3.88)	(3.58)	(13.5)			(6.73)
Z_1	15.1	59.3	318.3			121.4
m1	-6.4		-0.81		5.86	1.06
m2	0.64		3.40		5.51	1.01
R^2	0.066	0.155	0.315			
SEE		0.042	0.048	0.042	0.050	0.044

B) Dependent variable is $(I/K^B)_{it}$ Standard deviation = 0.095

	(i) IV Differences	(ii) Within Groups	(iii) OLS Levels	(iv) Within Groups	(v) OLS Levels	(vi) GLS
$Q^B{}_{it}$	0.0033	0.0139	0.0097	0.0134	0.0174	0.0134
	(0.82)	(7.30)	(5.71)	(7.01)	(6.20)	(7.95)
$(I/K^B)_{i,t-1}$	0.1876	0.1634	0.5438	0.19*	0.19*	0.2580
	(3.96)	(5.16)	(18.5)			(9.02)
Z_1	16.3	87.1	476.2			170.7
m1	-7.2		-3.25		6.64	-0.48
m2	0.23		2.57		6.06	1.09
R^2	0.048	0.189	0.435			
SEE		0.061	0.071	0.061	0.078	0.063

C) Dependent variable is $(I/K^C)_{it}$ Standard deviation = 0.074

	(i) IV Differences	(ii) Within Groups	(iii) OLS Levels	(iv) Within Groups	(v) OLS Levels	(vi) GLS
$Q^C{}_{it}$	-0.0005	0.0085	0.0053	0.0079	0.0083	0.0074
	(-0.09)	(4.26)	(3.55)	(4.18)	(4.24)	(4.41)
$(I/K^C)_{i,t-1}$	0.2232	0.1651	0.5059	0.22*	0.22*	0.2646
	(4.75)	(5.31)	(15.0)			(9.15)
Z_1	22.5	70.5	313.5			154.1
m1	-8.4		-2.75		5.51	-0.20
m2	-0.15		2.76		5.22	0.51
R^2	0.094	0.151	0.335			
SEE		0.053	0.060	0.053	0.064	0.055

Notes to Table 1

1) All equations contain time dummies.
2) Asymptotic t-ratios are reported in parentheses, robust to heteroscedasticity of arbitrary form.
3) Both variables are instrumented in column (i). The instrument set is $(I/K)_{i,t-2}$, $(I/K)_{i,t-3}$, $\Delta(I/K)_{i,t-3}$, $Q_{i,t-1}$, $Q_{i,t-2}$, $Q_{i,t-3}$, $\Delta Q_{i,t-3}$, $(Y/K)_{i,t-1}$, $(Y/K)_{i,t-2}$, $(Y/K)_{i,t-3}$ and $\Delta(Y/K)_{i,t-3}$, where Y denotes real company sales.
4) An asterisk indicates that a coefficient has been imposed in columns (iv) and (v).
5) Z_1 is a Wald test of joint significance, asymptotically distributed as $\chi^2(2)$.
6) m1 and m2 are tests of first-order and second-order residual serial correlation, asymptotically distributed as N(0,1) and robust to heteroscedasticity of arbitrary form. See Arellano and Bond (1988).
7) R^2 is the reduced-form R^2 in column (i) and the usual R^2 in columns (ii) and (iii).
8) SEE is the equation standard error.
9) The sample contains 191 companies and 1809 observations.
10) Capital stock measures A, B and C have been used to construct the measures of the investment rate and Q used in panels A, B and C of this table. Note that the dependent variable is therefore different in each panel.

where the instrument set contains lagged values of Q, the investment rate and the sales–capital ratio. Again we find that Q is not significant, although the coefficient on the lagged dependent variable is well determined.

An alternative transformation that also removes the firm effects is to express each variable in deviation from firm-mean form, i.e.

$$\tilde{x}_{it} = x_{it} - \frac{1}{T_i} \sum_{s=1}^{T_i} x_{is} \quad ,$$

where T_i is the number of time series observations on firm i. The time-invariant α_i are annihilated by this procedure, and performing OLS on the transformed equation produces the standard "within-groups" or "covariance" estimator. Unfortunately in short samples this estimator is biased, when the explanatory variables are not strictly exogenous (i.e. when $\text{cov}(x_{it}, \epsilon_{is}) \neq 0$ for all t, s), essentially because the transformed error term ($\tilde{\epsilon}_{it}$) contains ϵ_{is} ($s=1....T_i$), each multiplied by $1/T_i$. Only when the number of time series observations is large does the influence of each period's error become negligible. The lagged dependent variable is clearly not strictly exogenous, and the within-groups estimate of the coefficient on this variable is biased downwards in short samples (see Nickell (1981)). This is what we find in column (ii), where we report the within-groups estimates of the investment model. Q is significant in this specification, although this estimate may also be biased.

A third way of estimating the model is simply to ignore the firm effects and estimate the equation in levels by OLS, which we report in column (iii). Here we find the coefficient on the lagged dependent variable to be seriously biased upwards, which is to be expected since lagged investment is positively correlated with the omitted firm effects.

A simple way round these problems is to impose a consistent estimate of the coefficient on the lagged dependent variable. In columns (iv) and (v) we impose the value of 0.22 obtained from the differenced instrumental variables estimator (column (i)). We then estimate the coefficient on Q using the within-groups and OLS estimators respectively, and obtain very similar results. In both cases Q is found to be highly significant. The similarity of the OLS estimate to the within-groups estimate suggests that Q is not significantly correlated with the firm effects. In this case both β and γ can be estimated consistently and more efficiently using a two-step generalised least squares estimator, although, strictly, consistency again depends on the time dimension of the sample. GLS estimates can be obtained by performing OLS on equation (6) pre-multiplied by any matrix that is proportional to $\Omega^{-1/2}$, where Ω is the variance matrix of the composite error term ($\alpha_i + \epsilon_{it}$) in (6). It can be shown (see Hsiao (1986) for example) that a suitable transformation is simply to express each variable in the form

$$x_{it}^{*} = x_{it} - (1-\hat{\theta})\frac{1}{T_i}\sum_{s=1}^{T_i} x_{is} \quad ,$$

where

$$\hat{\theta}^2 = \frac{\hat{\sigma}_\epsilon^2}{\hat{\sigma}_\epsilon^2 + T_i\hat{\sigma}_\alpha^2}$$

and $\hat{\sigma}_\epsilon^2$ is, a consistent estimate of var(ϵ_{it}) and $\hat{\sigma}_\alpha^2$ is a consistent estimate of var(α_i). We obtain $\hat{\sigma}_\epsilon^2$ from the residuals in column (iv), and ($\hat{\sigma}_\epsilon^2 + \hat{\sigma}_\alpha^2$) from the residuals in column (v), so $\hat{\theta}$ is readily constructed. Applying this procedure to the investment equation results in the estimates reported in column (vi). These GLS estimates are very close to those obtained previously in columns (iv) and (v), so it would appear that finite sample biases are not a great problem here.

These results are very much in line with those presented in Blundell, Bond, Devereux and Schiantarelli (1987), which is to be expected given the similarity of the data. The lagged dependent variable is highly significant, although its role in the model is difficult to rationalise theoretically. Whilst Q is significant in columns (ii) – (vi), the estimated coefficient of about 0.015 implies an implausibly slow adjustment process to changes in Q, with a half-life of about thirty years. In first differences the coefficient falls further and is insignificantly different from zero.

In panel B of Table 1 we repeat this exercise using measures of Q and the investment rate where the capital stock has been estimated using method B. Our results are very close to those obtained with capital stock measure A. Q remains insignificant in first differences, and has an estimated coefficient of only around 0.013 in levels. In both cases the lagged dependent variable remains strongly significant. Despite the considerable differences in the data shown in Figures 2 and 3, the coefficient estimates in the investment equation are not highly sensitive to whether measures A or B of the capital stock are used.

In panel C we report estimates of the model using measure C of the capital stock to construct the data. Here the estimated coefficient on Q falls to only about 0.008, but otherwise the pattern of results is again similar.

It would appear from these results that econometric problems associated with measurement errors in the Q investment equations of panel A are not reduced when alternative estimates of the replacement value of net capital are used to construct the data. Making explicit allowance within the perpetual inventory approach for unusual obsolescence resulting from the first oil price shock has very little effect on our coefficient estimates. Following the approach suggested by Mairesse and Dormont (1985) seems to increase measurement error problems in this application. In all cases the familiar problems with the Q model are reproduced. On the positive side, this suggests that we do no worse by using the basic perpetual inventory method to construct capital stock estimates than we do with more complex alternatives. Since our measure A of the capital stock is closest in spirit to that found in national accounts, this may offer some reassurance to researchers working with aggregate data. On the other hand the reasons for the empirical weakness of the Q approach would appear to lie elsewhere.

One suggestion made above was that capital stock measure B may be less subject to error in the later part of our sample period, whether or not it deals successfully with the period 1974/5. Essentially this was because measure B exploits data on the replacement value of net fixed assets available in UK company accounts in the early 1980s. If this is the case we might expect data constructed using method B to perform better than those using method A over this more recent period. In Table 2 we re-estimate the Q investment equation using only the years 1979–84[5]. In fact we find little support for this hypothesis. In both cases the coefficient on the lagged investment rate falls, but it remains significantly different from zero. In both cases the coefficient on Q increases a little, but this improvement is less pronounced using measure B. Accountants' estimates of the replacement value of net capital perform no better than our own estimates in this context.

4. Conclusion

This paper has examined the degree to which the poor empirical results commonly found in estimating Q models of investment can be traced to measurement problems in the replacement value of the capital stock. The examination was undertaken by comparing the econometric results of a Q investment equation using 3 different measures of the capital stock. Two methods of estimating the capital stock were tested against the commonly used perpetual

TABLE 2 Q INVESTMENT EQUATIONS, 1979–84

GLS estimates

Dependent variable is $(I/K)_{it}$

	Capital Stock Measure A	Capital Stock Measure B
Q_{it}	0.0189	0.0140
	(5.47)	(4.61)
$(I/K)_{i,t-1}$	0.1335	0.1789
	(3.06)	(4.18)
Z_1	65.5	61.7
m1	-1.78	-2.78
m2	-0.30	-0.03
SEE	0.042	0.059
Standard deviation of dependent variable	0.059	0.086

Notes to Table 2

1) All equations contain time dummies.
2) Asymptotic t-ratios are reported in parentheses, robust to heteroscedasticity of arbitrary form.
3) Z_1 is a Wald test of joint significance, asymptotically distributed as $\chi^2(2)$.
4) m1 and m2 are tests of first-order and second-order residual serial correlation, asymptotically distributed as N(0,1) and robust to heteroscedasticity of arbitrary form. See Arellano and Bond (1988).
5) SEE is the equation standard error.
6) The sample contains 191 companies and 1045 observations.

inventory method. The first allowed for unusually high obsolescence of capital as a result of energy price increases in 1974/5. The second made use of accounting estimates of depreciation which varied across time and companies.

Neither of these two methods improved the empirical performance of the equation tested. In all three cases the equations had a very low (although well-determined) coefficient on tax adjusted Q, and a strongly significant lagged dependent variable. This suggests that the standard perpetual inventory method at least performs no worse than other methods of estimating the replacement value of the capital stock in Q models. In addition, it suggests that the reasons for the poor empirical performance of Q models may lie elsewhere - for example, in the strong assumptions underlying the empirical model concerning perfect competition, the efficiency of stock market valuations and capital markets, and the homogeneity of the capital stock.

References

Arellano. M. and S.R. Bond: "Some tests of specification for panel data: Monte Carlo evidence and an application to employment equations", mimeo, Institute for Fiscal Studies, 1988.

Baily, M.N.: "Productivity and the services of capital and labor". *Brookings Papers on Economic Activity* 1(1981), 1-65.

Blundell, R.W., S.R. Bond, M.P. Devereux and F. Schiantarelli: "Does Q matter for investment? Some evidence from a panel of UK companies". Institute for Fiscal Studies Working Paper No. 87/12, 1987.

Bond, S.R. and A. Daniels: "Estimating the replacement value of net fixed assets from UK company accounts", mimeo, Institute for Fiscal Studies, 1988.

Devereux, M.P.: "The IFS model of the UK corporation tax", IFS Working Paper No. 84, 1986.

Griffin, T.: "The stock of fixed assets in the United Kingdom: how to make best use of the statistics". *Economic Trends*, 276(10)(1976), 130-143.

Hayashi, F.: "Tobin's average q and marginal q: a neoclassical interpretation", *Econometrica*, 50(1982), 215-224.

Hayashi, F. and T. Inoue: "Implementing the Q theory of investment in micro data: Japanese manufacturing 1977-85", mimeo, Osaka University, 1987.

Hsaio, C.: *Analysis of panel data*, Cambridge: Cambridge University Press, 1986.

Hulten, C.R. and E.C. Wykopf: "The measurement of economic depreciation", in *Depreciation, inflation and the taxation of income from capital*, ed. by C.R. Hulten. Washington: Urban Institute, 1981.

King, M.A. and D. Fullerton: *The taxation of income from capital: a comparative study of the United States, the United Kingdom, Sweden and West Germany*. Chicago and London: University of Chicago Press, 1984.

Mairesse, J. and B. Dormont: "Labor and investment demand at the firm level: a comparison of French, German and US manufacturing, 1970–79", *European Economic Review*, 28(1985), 201-231.

Nickell, S.J.: "Biases in dynamic models with fixed effects", *Econometrica*, 49(1981), 1417-1426.

Poterba, J.M. and L.H. Summers: "Dividend taxation, corporate investment and 'Q'", *Journal of Public Economics*, 22(1983), 247-273.

Summers, L.H.: "Inflation, taxation and corporate investment: a q theory approach", *Brookings Papers on Economic Activity*, 1(1981), 67-140.

Wadhwani, S. and M. Wall: "The UK capital stock - new estimates of premature scrapping". *Oxford Review of Economic Policy* 2(1986), 44-55.

Data Appendix

This section describes the calculation of the principal variables used in the estimation. There are two data sources: company accounting records from Datastream and share price and related data from the London Share Price Database. We selected a sample of 191 of companies whose main activity, allocated by total sales, is in the manufacturing sector, which are available from both data sources and which have continuous data for at least 12 years. These companies are allocated to 9 industry groups as follows:

Group	Industry	SIC Classes	Number
1.	Metals and metal goods	21,22,31	24
2.	Other minerals and mineral products	23,24	21
3.	Chemicals and man made fibres	25,26	14
4.	Mechanical engineering	32,33	36
5.	Electrical and instrument engineering	34,37	17
6.	Motor vehicles and parts, other transport equipment	35,36	6
7.	Food, drink and tobacco	41,42	30
8.	Textiles, clothing, leather and footwear	43,44,45	17
9.	Other	46,47,48,49	26

Note: "Other" includes "timber and wooden furniture", "paper and publishing", "rubber and plastics", and "other manufacturing".

The variables are constructed as follows:

Investment ($p^I I$): Total new fixed assets, Datastream variable 435.

Share valuation (V): Using a monthly share price series for each company from the London Share Price Database, we calculate an average price over the three months preceding the accounting year. This average price is then multiplied by the total number of issued shares outstanding at the beginning of the year. We take an average price, rather than a beginning of the year price, in order to smooth out shocks to the share price.

Replacement cost of the capital stock ($p^I K$): As explained in the text, 3 methods are used to estimate the replacement cost value of the capital stock.

Method A: the replacement cost value of capital is calculated on a perpetual inventory method, i.e.:

$$p^I_{t+1} K_{t+1} = \frac{p^I_{t+1}}{p^I_t}(1-\delta) p^I_t K_t + p^I_{t+1} I_{t+1} \qquad (A1)$$

We use values of δ for UK manufacturing taken from King and Fullerton (1984) of 8.19% for plant and machinery and 2.5% for buildings. This method requires a valuation for an initial period. For this we assume equality of replacement cost and historic cost valuations of the capital stock in the first year of data, normally 1968. To reduce the impact of this rather arbitrary assumption on our results, we exclude the first three years (1968 to 1970) from the data used in estimation.

Investment is not split by asset in the data. In order to calculate the replacement cost from (A1) for each asset separately, we first use the split in the gross fixed asset figures to estimate the split in investment:

$$p^I_t I^P_t = p^I_t I_t \frac{(GFP_t - GFP_{t-1})}{(GFP_t - GFP_{t-1} + GFB_t - GFB_{t-1})} \qquad (A2)$$

$$p^I_t I^B_t = p^I_t I_t - p^I_t I^P_t \qquad (A3)$$

where $p^I_t I_t$ is the value of total investment, $p^I_t I^P_t$ and $p^I_t I^B_t$ the value of investment in plant and machinery and buildings respectively, and GFP_t and GFB_t are gross historic cost values of plant and machinery and buildings respectively, all in period t. The Datastream variables used for GFP_t and GFB_t are 328 and 327, respectively.

Method B: This also uses the perpetual inventory method. The initial value used is the first estimate of replacement cost provided in current cost accounts data (normally 1981). Where direct estimates are available from the data they are used. For earlier and later years, equation

(A1) is used. When calculating earlier years, this is transposed as:

$$p_t^I K_t = \frac{p_t^I}{p_{t+1}^I}(1-\delta)^{-1}\{p_{t+1}^I K_{t+1} - p_{t+1}^I I_{t+1}\} \tag{A4}$$

The same split of assets and rates of depreciation are used as for Method A. Method A estimates are used before 1974, and a weighted average (with weights depending on the company's accounting year end) of the method A and B approaches is used for 1974.

Method C: This makes use of the estimates of net historic cost fixed assets from the accounting data (Datastream variable 339), $NCHA_t$. These are adjusted as follows:

$$p_t^I K_t = \frac{p_t^I}{p_{t-A}^I} NCHA_t \tag{A5}$$

where A is the average age of net fixed assets. Using estimates from Bond and Daniels (1988), A was taken to be 3 years.

Other accouting variables: These are generally available directly from the data. The stock of debt at the end of the previous year (including any loan with repayment due in excess of one year (Datastream 321)) is used as an empirical proxy for H_t (see equation (2)).

Price indices: The output prices (P_t) are implicit value-added price deflators for the 9 sub-sectors of manufacturing industry. They were constructed from current price GDP and constant price GDP figures published by industry in various Blue Books. The price of investment goods (P^I_t) was derived as a implicit price deflator for gross fixed investment by manufacturing industry, taken from data in Economic Trends Annual Supplement (1987).

Tax parameters: These were estimated taking into account periods of tax exhaustion. We do not follow the common practice of assuming tax parameters to be zero in periods in which the company does not pay tax. Rather the existence of loss carry forward provisions implies that any liabilities or allowances are postponed until the company is once more in a tax paying position.

Consider the "adjusted" corporation tax rate, denoted by τ_t^*. If the company does not pay any tax until period t+n, then we define τ_t^* as

$$\tau_t^* = \frac{\tau_t}{(1+r_t)^n} \tag{A6}$$

where τ_t is the statutory corporation tax rate. As can easily be seen, for n = 0 we revert to the case with no tax exhaustion, while as n tends to ∞, τ^* tends to 0. It is clearly necessary to know all future periods of tax exhaustion. These have been estimated from the model of the UK corporation tax described in Devereux (1986). Accounting data for periods beyond the last year of data were forecast using a simple autoregressive model (see Devereux (1986)). We assume that companies have perfect foresight with regard to their own accounting data, but always expect there to be no change in statutory tax rates.

To assess investment allowances it is necessary to consider 2 types of asset, plant and machinery, and industrial buildings[6]. Commercial buildings receive no allowance, and can therefore be ignored. Following calculations in Devereux (1986), we assume that industrial buildings constitute 65% of the total value of buildings. We first outline the relevant tax parameters as if there were no tax exhaustion, and then show how they need to be modified in its presence.

Plant and machinery receive a first year allowance, α^P, and, in subsequent years, a depreciation allowance, δ^P, on a reducing balance basis. We assume that firms do not expect any changes in statutory tax rates, and ignore government grants (which were relatively unimportant during the period considered). We also ignore lags in payment of tax. We discount future allowances at the current rate on British government consols, which is approximately a risk-free rate. In the absence of tax exhaustion, the present value of investment allowances on a unit investment in plant machinery in period t, np_t, is[7]

$$np_t = \tau_t \alpha_t^P + \frac{\tau_t \delta_t^P (1-\alpha_t^P)}{(1+r_t)} \sum_{j=0}^{\infty} \left\{ \frac{(1-\delta_t^P)}{(1+r_t)} \right\}^j$$

$$= \tau_t \alpha_t^P + \frac{\tau_t \delta_t^P (1-\alpha_t^P)}{(r_t + \delta_t^P)} \qquad (A7)$$

The value of allowances on past investment in plant and machinery that can be claimed in period t, again on the assumption of no tax exhaustion, is:

$$APT_t = \tau_t \delta_t^P \left\{ (1-\alpha_{t-1}^P) I_{t-1} + \sum_{j=0}^{\infty} \left[\prod_{k=1}^{j} (1-\delta_{t-k}^P) \right] (1-\alpha_{t-j-1}^P) I_{t-j-1} \right\} \qquad (A8)$$

Summing these values into the future, and discounting at rate r_t, the present value of investment allowances on past investment in plant and machinery is:

$$AP_t = APT_t \sum_{j=0}^{\infty} \left\{ \frac{(1-\delta_t^P)}{(1+r_t)} \right\}^j$$

$$= APT_t \frac{(1+r_t)}{r_t + \delta_t^P} \qquad (A9)$$

By contrast, industrial buildings receive a straight line depreciation allowance[8], δ^B , together with an additional allowance in the year of investment, α^B. Hence, a building only receives an allowance for a fixed number of years, T, where

$$\alpha^B + (1+T)\delta^B = 1$$

and so

$$T = \frac{(1-\alpha^B-\delta^B)}{\delta^B} \qquad (A10)$$

Following a similar approach to that for plant and machinery, it can be shown that the present value of the investment allowances on a unit investment in industrial buildings, on the assumption of no tax exhaustion, is:

$$nb_t = \tau_t \alpha_t^B + \tau_t \delta_t^B \frac{(1+r_t)^{T+1}-1}{r_t(1+r_t)^T} \qquad (A11)$$

Calculation of the present value of remaining allowances on past investment in industrial building is slightly more complex. Consider an investment in period s (s<t). The present value of remaining allowances on this investment at period t is

$$ABT_{st} = \sum_{j=0}^{s+T_s-t} \frac{\tau_t I_s \delta_s^B}{(1+r_t)^j} \qquad (A12)$$

The sum of all such allowances includes all periods, s, for which s + T_s > t. If we define the total of such periods as F_t, then the present value of investment allowances on past investment in industrial buildings is

$$AB_t = \sum_{k=1}^{F_t} ABT_{kt} \qquad (A13)$$

Now $A_t = AP_t + AB_t$ and $n_t = (np_t I^P{}_t + nb_t I^B{}_t)/I_t$.

To take account of tax exhaustion, suppose the company first pays tax in period t+n, and then not again until period t+n+m. We can then define the present value of investment allowances on a unit investment in plant and machinery to be

$$np_t^* = \frac{\tau_t \alpha_t^P}{(1+r_t)^n} + \frac{\tau_t \delta_t^P (1-\alpha_t^P)}{(1+r_t)^n} \sum_{j=0}^{n-1} (1-\delta_t^P)^j$$

$$+ \frac{\tau_t \delta_t^P (1-\alpha_t^P)(1-\delta_t^P)^n}{(1+r_t)^{n+m}} \sum_{k=0}^{m-1} (1-\delta_t^P)^k + \ldots \qquad (A14)$$

and in industrial buildings to be

$$nb_t^* = \frac{\tau_t \alpha_t^B}{(1+r_t)^n} + \frac{\tau_t \delta_t^B (n+1)}{(1+r_t)^n} + \frac{\tau_t \delta_t^B m}{(1+r_t)^{n+m}} + \ldots \quad (A15)$$

for $n+m \leq T_t$

Similarly, we can define the present value of allowances on past investment in plant and machinery to be

$$AP_t^* = \frac{APT_t}{(1+r_t)^n} \sum_{j=0}^{n} (1-\delta_t^P)^j + \frac{APT_t}{(1+r_t)^{n+m}} \sum_{k=n+1}^{m} (1-\delta_t^P)^k + \ldots \quad (A16)$$

The present value in year t of allowances on an investment in industrial buildings in year s is

$$ABT_{st}^* = \frac{(n+1)\tau_t I_s \delta_s^B}{(1+r_t)^n} + \frac{m \tau_t I_s \delta_s^B}{(1+r_t)^{n+m}} + \ldots \quad (A17)$$

for $n + m \leq T_s$

As before:

$$AB_t^* = \sum_{k=1}^{F_t} ABT_{st}^* \quad (A18)$$

End Notes

1. The authors are grateful for many helpful comments from Manuel Arellano, Richard Blundell, Fabio Schiantarelli and participants in the "Factors in Business Investment" conference in Berlin, September 1987. Errors remain their own. They acknowledge the financial support of the ESRC under project B00232207.

2. We do not consider the further adjustments that would be introduced by personal taxes.

3. Strictly, the value of H_t also changes in the presence of taxation due to the deductibility of nominal interest payments from the tax base.

4. Comparisons over time are not entirely meaningful, since the sample varies over time. However, comparisons between the different methods are valid.

5. Here we report only the GLS estimates (cf. Table 1, column (vi)). Although the sample period is shorter, comparison with other estimators again gave no indication of finite sample biases.

6. Motor vehicles were treated as if they were taxed in the same way as plant and machinery.

7. For ease of expostion we ignore expectations terms. We assume throughout that the firm expects statutory tax rates and the discount rate to remain unchanged.

8. δ^B remains fixed over the lifetime of the asset.

Taxation and Capital Formation in West German Industries: A Q-Theory Approach

Michael Funke*, Andreas Ryll and Dirk Willenbockel

Science Centre Berlin, Research Unit Labour Market and Employment

I. Introduction

The purpose of this paper is to examine empirically the determinants of industry investment in West Germany. The traditional neoclassical theory of investment developed by Jorgenson[1] among other economists does not provide a fully satisfactory explanation of the determinants of investment behaviour. The dissatisfaction is not based on the explanatory power of the estimated investment equations, rather the problems arise from the difficulty of interpreting the resulting lag structures. An alternative approach uses the so-called Q-theory of investment which is based on Tobin's Q-concept and explicitly incorporates the cost of adjustment literature. The question whether this hypothesis is a better approximation to the behaviour of real-world investment has major substantive implications for both macroeconomic and financial economic theory and practice. This paper extends research in the area of the Q-theory of investment. Moreover, our contribution differs in two important respects from earlier studies of the Q-theory of investment in West Germany. First, all existing studies analyse investment behaviour at an aggregate level, i.e. for total industry or total manufacturing. We have also computed investment equations for broad sectors but, in addition, our analysis covers sixteen major industrial sectors out of total manufacturing separately. Secondly, contrary to other existing studies we have used tax-adjusted marginal Q-measures instead of unadjusted average Q-values. The article is organized as follows. The next section describes the investment model. The third and the fourth section review the historical trends in sectoral investment rates and Q-values and present the econometric specification and

---

* The authors are grateful to Dr. E. Dinenis of the London Business School for many helpful discussions. Additionally, we would like to thank S. Wadewitz for her expert research assistence and W. Röger for helpful suggestions. Responsibility for errors is ours, of course. Financial support from DFG (Deutsche Forschungsgemeinschaft) is gratefully acknowledged.

model. The final section contains a summary of our findings. In appendix I, II, and III abbreviations and computation methods are given.

II. The Model of the Firm

In this section of the paper we develop a theoretical model of the determinants of private investment based on the Q-theory of investment.[2] The Q model proposes that investment spending rises directly with the ratio of the market value of business capital assets to the replacement value of those assets. This ratio, commonly known as "Q", compares the yield on new capital goods with the rates of return required by lenders of capital. In other words: managers can calculate the price they are willing to pay for an investment project after assessing its potential returns; this price is the demand price for capital goods. The cost of producing new capital goods determines the supply price. In equilibrium both prices are normally equal. Should a change in the business conditions create profitable investment opportunities, however, the Q ratio and therefore investment will increase. The demand and the supply price of capital will equate again when the profitable investment opportunities are exhausted. This story implies that the financial markets (stock markets) continually assess the prospective return offered by enterprises. In the following model we provide a rigorous foundation of the Q-theory of investment based on the optimization by stock companies in the presence of taxation. Following Hayashi (1982), Abel[3], Summers (1981) and Yoshikawa (1980) we assume that managers maximize the expected present value of the firm's equity. At t = 0, the value of the firm's shares V(o) is given by

$$V(o) = \int_0^\infty \theta \, Div(t) \, \mu(t) \, dt \qquad (1)$$

where

$$\mu(t) = \exp \left(\int_0^t - r(s)ds\right) \qquad (2)$$

and Div(t) = nominal dividends on shares, r = exogenous equity return (discount rate), θ = King's tax discrimination variable. In the context of our model θ is introduced to capture the different tax treatment of retentions vis-â-vis distributions in West Germany in a consistent manner (both under the split-rate system in force before 1977 and under the imputation system afterwards), while preserving a unique profit tax rate in the model equations.[4] Additionally we impose the transversality condition

$$\lim_{t \to \infty} V(t) \exp \int_0^t - r(s)ds = 0. \qquad (3)$$

In the presence of taxation and exogenous sources of funds dividends may be derived as after-tax profits less investment expenses plus new net debt. That is,

$$Div(t) = (1-\tau)(pF(K,N)-wN-iB)+\tau D(t)+B^n(t)-\varepsilon B(t)-(1-k)pI(t) \qquad (4)$$

where K and N refer to factor inputs, p is the price level, $F(\cdot)$ is the production function, τ is the corporate tax rate on retained earnings, k captures investment grants (net of investment tax), D represents the amount of depreciation allowances, i is the interest rate on debt B, ε is the amortization rate on debt, $B^n(t)$ is new debt and I represents real investment. The capital accumulation constraint is given by

$$\dot{K}(t) = I - \delta K \qquad (5)$$

where the dot over K indicates the time derivative, I and K give gross investment and the capital stock, respectively, and δ is the depreciation rate.

As is well known, there must be some positive installation or adjustment cost because otherwise the optimal level of investment remains indeterminate.[5] It is commonly assumed that installation costs take the form of output losses or wasted capital. We model the assumption that part of investment (I) is wasted in the installation process by using a so-called installation function $\Psi(I,K)$[6], which relates uninstalled investment (I) to gross additions to the installed capital stock (Ψ). Thus, instead of (5) we have

$$\dot{K} = \Psi(I,K,t) - \delta K(t).\ \Psi_I > 0;\ \Psi_{II} < 0;\ \Psi_K > 0 \qquad (6)$$

If we assume that total debt is exogenous and therefore independent of investment decisions by the firm,[7] the optimizing problem reduces after some simple manipulations to [8]

$$V(0) = \int_0^\infty \theta[(1-\tau)(pF(\cdot)-wN) - (1-k-z)pI]\ \mu(t)dt + \theta A(o) - \theta B(o) \qquad (7)$$

where $A(o) = \int_0^\infty \tau(t)\ [\int_{-\infty}^0 d(t-v,v)pI(v)dv]\ \mu(t)dt$

$$z(t) = \int_0^\infty \tau(t+s)\, d(s,t)\, \mu(s) ds$$

The term A(o) represents the present value of current and future tax deductions which are attributable to *past* investment and which is therefore irrelevant for investment decisions from t = 0 onward and can thus be ignored. The variable z(t) represents the present value of the tax bill saving due to depreciation deductions on a D-Mark of investment installed at time t. The optimization problem consists in maximizing equation (7) subject to the capital accumulation constraint (6). The optimality conditions from the Hamiltonian are:

$$pF_N = w \tag{8}$$

$$\theta(1 - k - z)p = \Psi_I(I,K,t) \cdot \lambda \tag{9}$$

$$\dot{\lambda} = \lambda(r + \delta - \Psi_K) - \theta(1 - \tau)\, pF_K \tag{10}$$

$$\lim_{t \to \infty} \lambda(t)\, K(t)\, \mu(t) = 0 \tag{11}$$

where λ is the shadow price of capital. Equation (8) gives the familiar marginal productivity condition for the use of labor. Equation (9) is the key equation because it yields the investment-Q relationship. It demonstrates that the rate of gross investment is chosen so as to equate the marginal cost of investment with the shadow price of installed capital. Equation (10) determines the dynamic behaviour of the shadow price λ. Under the assumption of linear-homogeneity of $\Psi(\cdot)$ equation (9) can be inverted to give the rate of investment.

$$I/K = f\left(\frac{\lambda/p}{\theta(1-k-z)}\right) \tag{12}$$

$$I/K = f\left(\frac{q_m}{\theta(1-k-z)}\right) \tag{13}$$

$$I/K = f(Q_{m,d}) \tag{14}$$

where $q_m = \lambda/p$ = marginal Q

$Q_{m,d} = q_m/[(1-z-k)\theta]$ = tax-adjusted marginal Q.

In other words: the optimizing model leads to an investment equation linking investment to marginal Q and the tax parameters.

Furthermore we can use the above model structure to derive an exact relationship between average Q and marginal Q and therefore a relationship between the investment rate and the observable average Q value (Q_a).[9]

Proposition: If the production function and the adjustment cost function are linear homogeneous and if the firm is a price taker, then:[10]

$$Q_{m,d} = \frac{V(o)}{\theta(1-k-z)pK} + \frac{B(o)-A(o)}{(1-k-z)pK} \tag{15}$$

$$\frac{I}{K} = f\left(\frac{V(o)}{\theta(1-k-z)pK} + \frac{B(o)-A(o)}{(1-k-z)pK}\right) \tag{16}$$

The proposition presented above is essentially an extension of the results derived by Hayashi. This can easily be demonstrated because it follows directly that in the special case of $\theta = 1$ equation (16) reduces to Hayashi's equation (13") (Hayashi, F. 1982: 220).

While equation (14) is our basic empirical q model, it is by no means the only possible rigorously derived specification. Hayashi (1982: 216) has pointed out that if installation costs take the form of lost output and the production function is *non-separable*, then the investment behaviour will depend on Q_m as well as on the real wage (w/p). This alternative investment function can be written in the form:

$$I/K = f\,(Q_{m,d},\ (w/p)). \tag{17}$$

III. Historical Trends in Sectoral Investment Rates and Q Values

Up to this point we have developed a model relating tax-adjusted marginal Q values to the present investment rate of the firm. Before estimating the resulting investment equations for 19 industry subgroups we first briefly review the investment rate[11] performance within the different industries in the sample. Figure I and figure II graph the various gross and net rates of investment for *total industry* and *manufacturing industry* for the period 1960 to 1982. For each sector we have broken down investment by two asset types - total investment and investment in plant and equipment.[12] A visual inspection of the graphs shows that for total industry and manufacturing industry all investment rates have declined substantially over the sample period. All time series portray a consistent pattern. While in the period from 1960 to 1970 only a slight reduction has occured, all investment rates have declined considerably more rapidly during the final subperiod 1970-1982. Even more

Figure I and II

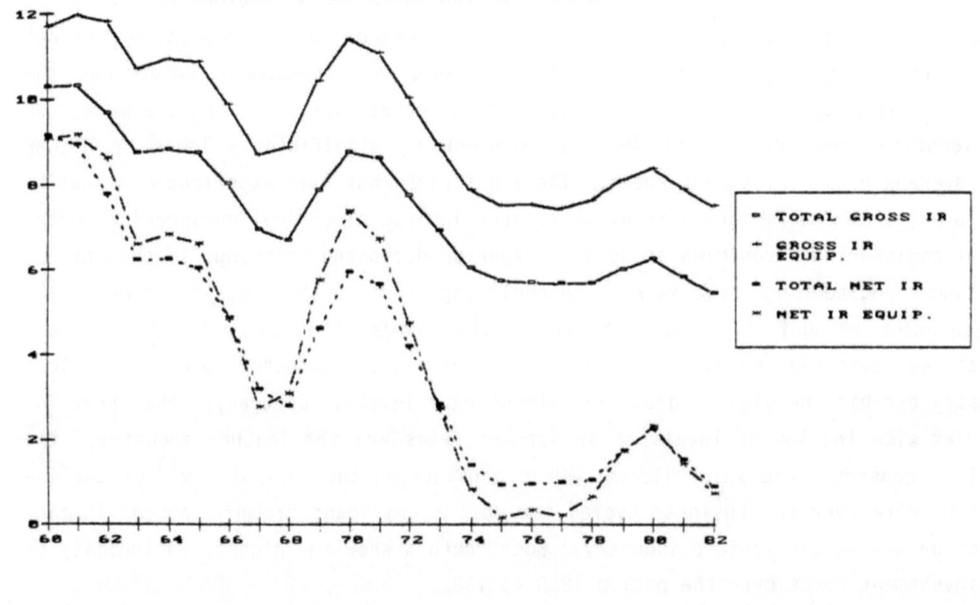

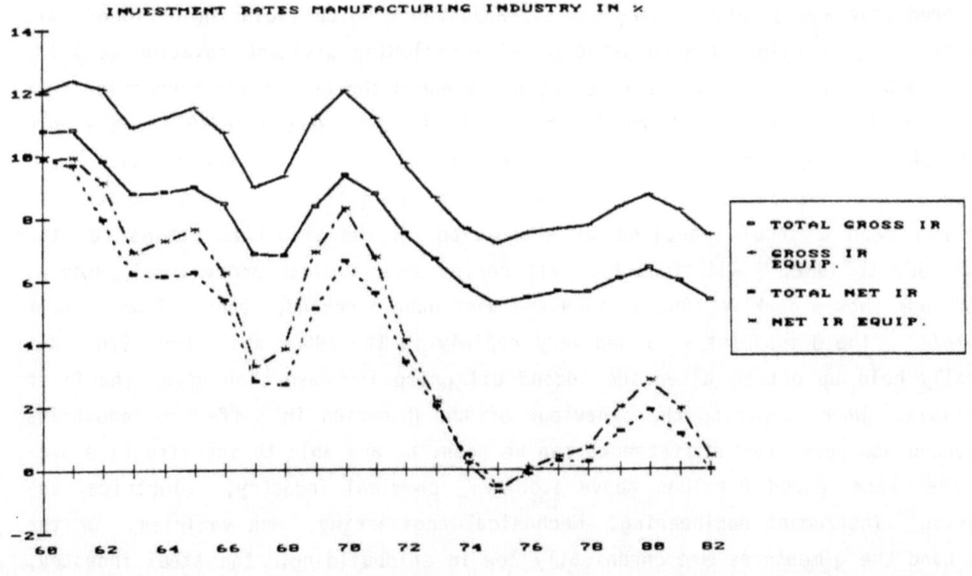

dramatic than the decline in gross investment rates was the decline in net investment rates. This decline appears to be the result of the shift from net investment to replacement investment. The increasing share of replacement investment is probably the natural outcome of growth in capital stock and the shift in investment from long-lived to short-lived capital goods, e.g. from structures to plant and equipment which have shorter lives and therefore higher replacement rates. This pattern of net investment rates plays an important role because *net* investment determines the economy's capacity to increase its productive potential over time. Furthermore, we can identify from our dataset sectors experiencing significantly lower or higher than average gross investment rates. The industries that have experienced a smaller than average *decline* in gross investment rates include electrical engineering, mechanical engineering, vehicles and paper, paper products and printing. In the case of instrument engineering there is even no declining trend in the gross investment rate in equipment recognizable. When comparing the *levels* of investment rates between industries again the engineering industries, the vehicle industry, and the chemical industry exhibit the highest gross investment rate levels. Contrary, the three industries with the lowest levels of investment rates are the leather industry, the textile industry, and shipbuilding. When considering the variability[13] of the investment rate over the business cycles the most significant features appear in construction and in the vehicle industry. Both sectors show the highest variability of the investment rates over the period 1960 to 1982.

In the following the Q measures are reviewed briefly. Our first general observation on figure III and figure IV is that the coincidence of movements in the Q measures considered (average Q value, marginal tax-adjusted Q value excluding dividend taxation ($\theta = 1$), marginal tax-adjusted Q value including dividend taxation ($\theta \neq 1$)) appears to be high. Despite the absolute differences the prevailing impression from the figures is one of overall similarity. Only in the years 1976/1977 the stable relationship between both marginal Q measures breaks down due to the increase in θ because of the corporate tax reform in West Germany in 1977. In most industries[14] there has been a secular decline or stagnation in the valuation ratios for the period 1960 to 1982.[15] Additionally, all series show a clear procyclical pattern. All of them show a peak in 1968 or 1969 and most have a second, though lower, peak in 1976/77. The Q measures declined very rapidly in the 1960s and after 1970, but generally held up better after the second oil price increase than after the first oil crisis. When comparing the behaviour of the Q series in different industries pronounced and persistent differences can be seen. We are able to identify five sectors that experienced Q values above average: chemical industry, electrical engineering, instrument engineering, mechanical engineering, and vehicles. On the other hand the Q measures are chronically low in shipbuilding, the steel industry, woodworking and woodprocessing, and textiles.[16] Over time this distribution seems to

Figure III and IV

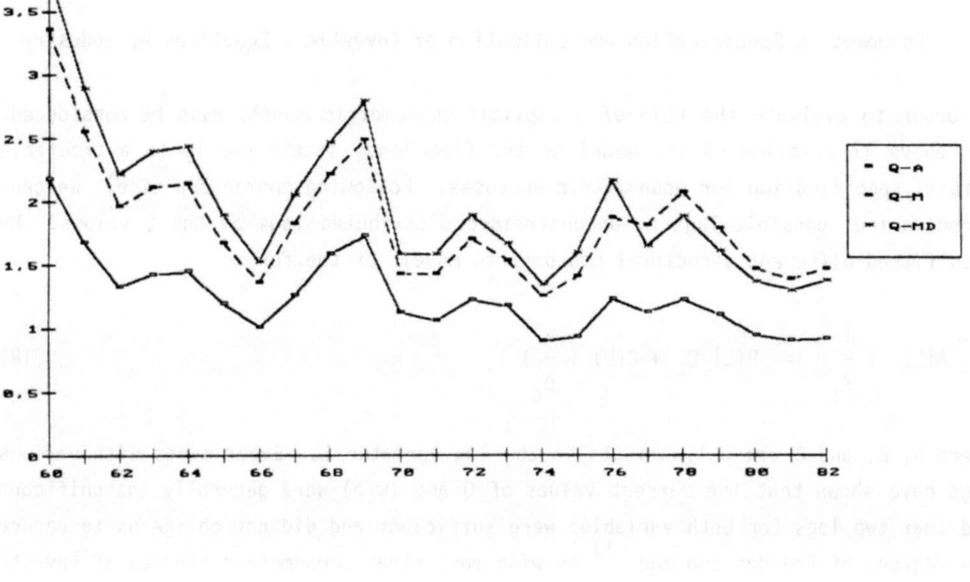

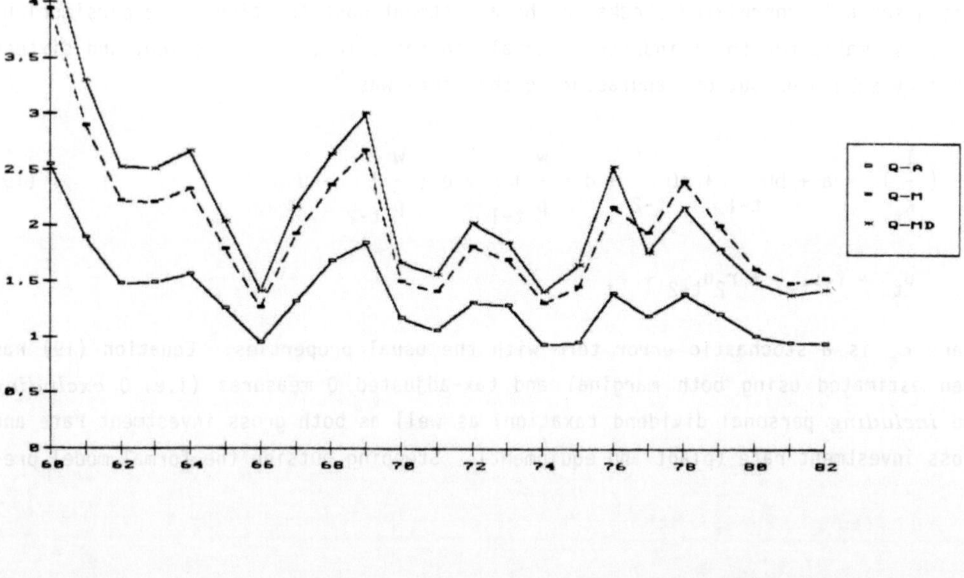

be relatively stable although a remarkable decline in the Q measures of the engineering industries in the 1970s can be identified. From a dynamic perspective, it seems that industries which are characterized by high R & D expenditures and rapid process and product innovations can maintain better valuation ratios and offer more attractive investment opportunities.

IV. Econometric Specification and Estimation of Investment Equations by Industry

In order to evaluate the role of Q explicit econometric models must be considered. The above formulation of the model of the firm lends itself easily to a tractable initial specification for econometric purposes. Following common practice, we captured various possible lags by unconstrained distributed lags on the Q values. We then fitted different structural econometric models of the form

$$A(L)\left(\frac{I}{K_t}\right) = B(L)\,Q_t + C(L)\left(\frac{w}{p_t}\right) \tag{18}$$

where A, B, and C are polynominals in the lag operator L. Experiments with various lags have shown that the current values of Q and (w/p) were generally insignificant and that two lags for both variables were sufficient and did not oblige us to reduce the degrees of freedom too much.[17] As with most other econometric studies of investment behaviour, we found strong evidence of autocorrelation in the unadjusted estimates of the model in equation (18). Because of the lag length of two years it is not surprising that adjustment for both first-order and second-order autoregressive time-series pattern is significant in all fitted equations.[18] From a theoretical point of view this serial correlation is the error process can be justified by assuming serially correlated shocks to the adjustment cost function. The parsimonious model estimated for total industry, total manufacturing, construction, and sixteen industry subgroups out of manufacturing therefore was[19]

$$\left(\frac{I}{K_t}\right) = a + bQ_{t-1} + cQ_{t-2} + d\left(\frac{w}{p}\right)_{t-1} + e\left(\frac{w}{p}\right)_{t-2} + u_t \tag{19}$$

$$u_t = r_1 u_{t-1} + r_2 u_{t-2} + \varepsilon_t$$

where ε_t is a stochastic error term with the usual properties. Equation (19) has been estimated using both marginal and tax-adjusted Q measures (i.e. Q *excluding* and *including* personal dividend taxation) as well as both gross investment rate and gross investment rate (plant and equipment). Stepping outside the formal model presented in section II of the paper we finally have tested the existence and significance of accelerator or demand effects besides Q, i.e., we tested whether demand variables affect the investment rate directly and not indirectly through Q. While demand effects are accounted for in the Q variable, it is possible that Q is unable to capture the full extent of this influence. We have therefore included this "over-identifying restriction" by adding the growth rate of gross value added in constant (1980) prices by industry to equation (19).

All equations were fitted to annual data for the period 1965-1982.[20] The results are presented in full detail in table I and II below. All three equations were fitted using both marginal tax-adjusted Q series (including and excluding dividend taxation). Furthermore we compiled separate results for total gross investment and equipment investment.

Note first the diagnostic statistic. The explanatory power of the equations is high and the Durbin Watson statistic indicates the absence of serial correlation in most equations.[21] Nearly all of the Q values have the anticipated signs and are significant. A comparison of the results for different industries reveals that there are four sectors where the coefficients of the lagged Q values fade into insignificance in all equations: these are construction, iron and steel, shipbuilding, and plastic and rubber products.[22] In the case of iron and steel and shipbuilding these abnormalities in investment behaviour can probably be explained by the fact that both industries are heavily subsidized and are largely run as an employment program. Another general tendency seems to be that the strongest relationship between I/K and Q appears in the major industrial sectors. This may be due to the fact that the number of companies in the minor sectors of our sample is rather small thereby causing a bias in their Q values. All of the results are confirmed when the gross investment rate in equipment is used as a dependent variable. Although the correlation, judged by standards of R^2, is normally higher for equipment investment the equations for both investment rates correspond to orders of magnitude and significance which are roughly the same. The prevailing impression is therefore one of overall similarity.[23] On the whole, these rather unequivocal results are a bit surprising since the stock market in West Germany plays a comparatively minor role in financing investment compared to the U.K. or the U.S. This might suggest that the link between Q and investment arises less because of the direct part played by the stock market than because the sentiment underlying the stock market is the same which favors investment activity. In addition to our "equation" containing only lagged Q values we have estimated equations testing for the significance of real wages in a Neokeynesian investment equation. However, this modification receives little support from the data. The coefficients on the real wage are generally insignificant and often wrong signed.[24]

Table I: *Dependent variable:* *total gross investment rate (in %)*
Q-measure: *marginal tax-adjusted Q value* <u>*including*</u> *dividend taxation*
Sample period: *1965-1982*

	CONSTANT	Q_{t-1}	Q_{t-2}	WR_{t-1}	WR_{t-2}	GY	RHO_1	RHO_2	R^2	DW
	4,85	0,45	0,50				1,33	-0,56	0,91	1,35
	(6,1)	(2,0)	(2,4)				(6,2)	(-2,6)		
WG	8,63	0,37	0,15	0,19	-0,44		1,22	-0,85	0,96	1,47
	(9,1)	(1,8)	(0,8)	(0,7)	(-1,6)		(7,9)	(-5,5)		
	8,89	0,20	0,07	0,15	-0,39	0,03	1,21	-0,79	0,96	1,53
	(8,4)	(0,7)	(0,3)	(0,5)	(-1,3)	(0,8)	(7,8)	(-4,5)		
	3,98	0,69	0,61				1,30	-0,51	0,91	1,52
	(4,5)	(3,4)	(3,0)				(5,8)	(-2,3)		
VG	7,71	0,60	0,30	0,23	-0,48		1,20	-0,78	0,95	1,64
	(7,0)	(3,0)	(1,3)	(0,6)	(-1,3)		(6,4)	(-4,0)		
	7,72	0,18	0,13	0,03	-0,21	0,08	1,27	-0,72	0,97	1,60
	(6,3)	(0,7)	(0,7)	(0,1)	(-0,6)	(2,6)	(9,2)	(-4,9)		
	7,71	0,32	0,63				1,21	-0,72	0,76	2,14
	(7,9)	(0,6)	(1,7)				(5,9)	(-3,5)		
BG										
	(*) 8,49	-0,44	-0,02			0,22	0,71	-0,11	0,81	1,78
	(8,0)	(-0,6)	(-0,03)			(2,2)	(2,4)	(-0,4)		
	1,61	0,76	0,74				0,91	-0,41	0,87	2,38
	(1,4)	(3,4)	(3,6)				(4,2)	(-2,0)		
CH	7,85	0,35	0,23	-0,26	0,02		0,46	-0,61	0,91	1,97
	(4,4)	(1,5)	(0,9)	(-0,8)	(0,05)		(2,0)	(-3,0)		
	(*) 313,58	0,53	0,78	-0,50	0,27	0,04	1,23	-0,23	0,83	2,00
	(0,002)	(1,7)	(2,8)	(-1,0)	(0,5)	(0,7)	(3,1)	(-0,8)		
	9,83	-0,31	-0,06				1,14	-0,37	0,76	1,74
	(3,4)	(-0,6)	(-0,1)				(4,4)	(-1,4)		
KU	16,99	-0,38	-0,04	0,17	-0,82		0,94	-0,64	0,86	2,01
	(10,1)	(-0,7)	(-0,1)	(0,2)	(-0,7)		(3,9)	(-2,6)		
	12,12	-0,18	0,49	-1,15	0,75	0,14	0,95	-0,49	0,96	2,56
	(5,4)	(-0,6)	(1,4)	(-1,6)	(1,0)	(5,0)	(6,9)	(-3,5)		
	3,27	1,16	0,56				1,17	-0,61	0,82	1,71
	(2,1)	(2,8)	(1,3)				(5,7)	(-3,1)		
SE	7,76	0,86	0,28	-0,16	-0,14		1,05	-0,73	0,88	2,13
	(3,3)	(2,1)	(0,6)	(-0,3)	(-0,2)		(5,5)	(-3,8)		
	7,14	0,69	0,22	-0,34	0,12	0,08	1,03	-0,58	0,92	2,48
	(3,1)	(1,8)	(0,6)	(-0,6)	(0,2)	(2,1)	(6,1)	(-3,2)		
	1,56	0,25	2,73				0,97	-0,37	0,66	1,50
	(0,5)	(0,1)	(1,6)				(3,7)	(-1,4)		
ES	6,25	-0,03	1,80	$-7,6 \cdot 10^{-3}$	-0,24		0,92	-0,62	0,76	1,60
	(1,7)	(-0,01)	(1,0)	(-0,005)	(-1,5)		(3,9)	(-2,6)		
	(*) 673,37	$1,8 \cdot 10^{-3}$	3,33	0,34	$-5,1 \cdot 10^{-3}$	0,01	1,17	-0,17	0,58	1,76
	(0,004)	(0,001)	(2,0)	(0,6)	(-0,01)	(0,2)	(3,4)	(-0,5)		
	2,43	0,22	1,72				0,26	-0,21	0,66	2,03
	(2,5)	(0,4)	(3,1)				(1,0)	(-0,9)		
NE	4,10	0,04	1,54	-0,22	0,15		0,25	-0,09	0,67	1,99
	(1,5)	(0,1)	(2,4)	(-0,5)	(0,4)		(0,8)	(-0,3)		
	(*) 3,79	$-7,4 \cdot 10^{-4}$	1,59	-0,20	0,15	0,02	0,25	-0,11	0,68	1,95
	(1,2)	(0,001)	(2,3)	(-0,4)	(0,3)	(0,2)	(0,8)	(-0,4)		
	4,21	0,88	0,63				0,78	-0,53	0,59	2,09
	(3,6)	(1,6)	(1,2)				(3,3)	(-2,3)		
SL	(*) 5,14	0,75	0,48	-0,02	-0,02		0,79	-0,53	0,59	2,11
	(1,4)	(1,0)	(0,6)	(-0,1)	(-0,05)		(3,0)	(-1,9)		
	(*) 5,25	0,73	0,46	-0,05	$-6,3 \cdot 10^{-3}$	$4,1 \cdot 10^{-3}$	0,79	-0,52	0,59	2,10
	(1,4)	(0,9)	(0,6)	(-0,1)	(0,01)	(0,1)	(2,8)	(-1,6)		
	3,52	0,89	1,27				1,22	-0,76	0,92	1,86
	(6,0)	(2,6)	(5,1)				(6,4)	(-4,1)		
NA	1,53	1,26	1,42	0,12	-0,03		1,22	-0,77	0,93	1,88
	(0,6)	(2,4)	(3,5)	(0,7)	(-0,2)		(5,7)	(-3,5)		
	(*) 466,32	1,66	1,57	0,50	0,26	0,02	1,55	-0,55	0,88	1,29
	(0,01)	(2,3)	(1,8)	(1,3)	(0,1)	(0,5)	(5,5)	(-2,0)		

(*) Convergence not achieved; 't'-statistic in parantheses

Table I: *Dependent variable:* *total gross investment rate (in %)*
Q-measure: *marginal tax-adjusted Q value including dividend taxation*
Sample period: *1965-1982*

	CONSTANT	Q_{t-1}	Q_{t-2}	WR_{t-1}	WR_{t-2}	GY	RHO_1	RHO_2	R^2	DW
	7,58	0,17	0,14				1,23	-0,62	0,79	1,71
	(8,5)	(1,9)	(1,6)				(6,0)	(-3,0)		
SF	9,03	0,15	0,17	-0,25	0,15		1,22	-0,67	0,80	1,65
	(3,3)	(1,5)	(1,5)	(-0,8)	(0,5)		(5,6)	(-2,8)		
	(*) 23,25	0,05	0,06	-0,25	-0,60	0,11	1,37	-0,30	0,81	1,84
	(2,3)	(0,5)	(0,5)	(-0,5)	(-1,3)	(2,4)	(6,0)	(-1,3)		
	3,77	-0,04	1,68				0,90	-0,54	0,64	1,76
	(2,5)	(-0,04)	(1,4)				(4,2)	(-2,5)		
SB	3,27	0,30	1,94	-0,29	0,30		0,88	-0,60	0,66	1,71
	(0,8)	(0,2)	(1,3)	(-0,7)	(0,8)		(3,9)	(-2,7)		
	(*) 697,04	0,23	2,71	0,41	-0,02	0,03	1,05	-0,05	0,42	1,84
	(0,002)	(0,1)	(1,2)	(0,6)	(-0,03)	(0,6)	(3,6)	(-0,1)		
	3,76	0,83	0,74				1,06	-0,48	0,82	2,01
	(3,3)	(2,9)	(2,9)				(4,4)	(-1,9)		
ET	3,48	0,84	0,76	-0,12	0,15		1,06	-0,48	0,82	2,02
	(1,3)	(2,6)	(2,5)	(-0,2)	(0,3)		(4,0)	(-1,8)		
	5,22	0,18	0,42	-0,03	0,03	0,10	0,91	0,37	0,90	2,73
	(1,9)	(0,5)	(1,4)	(-0,1)	(0,1)	(2,9)	(5,0)	(-1,9)		
	5,90	0,32	0,35				1,18	-0,41	0,60	1,82
	(5,4)	(1,8)	(1,9)				(4,4)	(-1,5)		
FM	8,61	0,40	0,34	0,35	-0,61		1,31	-0,59	0,74	2,27
	(3,4)	(2,4)	(2,1)	(1,3)	(-2,5)		(5,0)	(-2,1)		
	(*) 8,91	-0,24	-0,10	-0,05	-0,07	0,10	1,00	-0,67	0,82	2,30
	(7,5)	(-1,3)	(-0,7)	(-0,2)	(-0,2)	(2,7)	(4,7)	(-3,3)		
	3,92	1,12	0,47				1,16	-0,44	0,90	1,85
	(4,0)	(3,9)	(1,5)				(4,8)	(-2,0)		
EM	8,92	0,78	0,12	0,36	-0,70		1,21	-0,82	0,97	1,87
	(8,2)	(4,1)	(0,6)	(2,3)	(-4,4)		(8,3)	(-5,6)		
	9,02	0,74	0,07	0,34	-0,68	$7,7 \cdot 10^{-3}$	1,20	-0,79	0,97	1,92
	(7,4)	(2,9)	(0,3)	(1,9)	(-3,7)	(0,3)	(7,6)	(-4,3)		
	4,65	0,75	0,43				1,24	-0,47	0,82	1,49
	(3,5)	(2,1)	(1,2)				(4,9)	(-1,7)		
HO	10,04	0,49	0,23	-0,21	-0,16		1,11	-0,56	0,85	1,68
	(4,0)	(1,4)	(0,6)	(-0,6)	(-0,4)		(4,4)	(-2,1)		
	7,42	0,10	-0,05	-0,31	0,19	0,17	0,65	-0,20	0,97	2,17
	(5,0)	(0,5)	(-0,2)	(-1,3)	(0,8)	(6,4)	(5,1)	(-1,5)		
	5,05	0,68	0,98				1,27	-0,56	0,77	1,79
	(4,7)	(1,5)	(2,5)				(5,5)	(-2,4)		
PD	8,34	0,79	0,64	0,26	-0,53		1,19	-0,59	0,82	2,26
	(3,6)	(1,6)	(1,5)	(0,8)	(-1,5)		(4,9)	(-2,4)		
	6,92	0,47	0,62	-0,05	-0,09	0,09	1,15	-0,51	0,87	2,68
	(2,6)	(1,0)	(1,6)	(-0,1)	(-0,2)	(2,0)	(5,5)	(-2,4)		
	0,97	0,44	0,64				0,81	0,04	0,87	1,98
	(0,8)	(2,2)	(3,2)				(3,7)	(0,2)		
LG	5,64	-0,26	1,11	0,27	-0,61		-0,81	-0,27	0,92	2,14
	(16,9)	(-1,8)	(8,0)	(1,4)	(-3,5)		(-2,8)	(-1,1)		
	-4,65	0,26	0,57	0,43	0,01	0,04	0,72	0,19	0,91	2,00
	(-0,8)	(1,2)	(3,1)	(1,8)	(0,04)	(1,3)	(3,2)	(0,9)		
	2,30	0,53	0,37				1,12	-0,22	0,84	1,90
	(1,1)	(1,7)	(1,3)				(4,1)	(-0,8)		
TG	7,18	0,40	0,26	-0,14	-0,26		0,82	-0,40	0,89	1,83
	(6,8)	(1,2)	(0,8)	(-0,3)	(-0,6)		(2,9)	(-1,5)		
	(*) 14,42	0,40	0,34	-0,39	$7,2 \cdot 10^{-3}$	0,10	0,93	0,06	0,94	0,71
	(0,1)	(1,9)	(1,6)	(-1,0)	(0,02)	(3,6)	(4,2)	(0,3)		

(*) Convergence not achieved; 't'-statistic in parantheses

Table II: *Dependent variable:* *total gross investment rate (in %)*
Q-measure: *marginal tax-adjusted Q value* *<u>excluding</u>* *dividend taxation*
Sample period: *1965-1982*

	CONSTANT	Q_{t-1}	Q_{t-2}	WR_{t-1}	WR_{t-2}	GY	RHO_1	RHO_2	R^2	DW
	4,53	0,50	0,64				1,35	-0,54	0,91	1,32
	(5,0)	(2,0)	(2,6)				(6,3)	(-2,5)		
WG	8,53	0,44	0,26	0,14	-0,40		1,20	-0,84	0,96	1,44
	(9,5)	(1,8)	(1,1)	(0,6)	(-1,6)		(7,7)	(-5,3)		
	8,75	0,27	0,18	0,11	-0,35	0,02	1,20	-0,79	0,96	1,52
	(8,7)	(0,8)	(0,7)	(0,4)	(-1,3)	(0,7)	(7,6)	(-4,4)		
	3,50	0,81	0,76				1,33	-0,49	0,91	1,48
	(3,3)	(3,4)	(3,3)				(5,9)	(-2,2)		
VG	7,65	0,77	0,39	0,24	-0,52		1,16	-0,78	0,95	1,51
	(7,7)	(3,1)	(1,5)	(0,7)	(-1,5)		(6,2)	(-4,0)		
	7,74	0,21	0,21	$-5,3\cdot 10^{-3}$	-0,19	0,08	1,25	-0,71	0,97	1,58
	(6,5)	(0,6)	(0,9)	(-0,02)	(-0,5)	(2,3)	(8,7)	(-4,6)		
	7,66	0,32	0,75				1,21	-0,71	0,77	2,12
	(7,5)	(0,6)	(1,7)				(5,9)	(-3,5)		
BG										
	(*) 8,28	-0,21	0,24			0,20	0,77	-0,23	0,80	1,81
	(8,2)	(-0,3)	(0,4)			(1,9)	(2,6)	(-0,8)		
	1,30	0,83	0,87				1,0	-0,40	0,85	2,34
	(0,9)	(2,9)	(3,3)				(4,7)	(-1,9)		
CH	7,80	0,43	0,31	-0,30	0,03		0,39	-0,61	0,92	2,16
	(5,5)	(1,7)	(1,1)	(-1,0)	(0,1)		(1,8)	(-3,1)		
	(*) 266,82	0,55	0,84	-0,44	0,12	0,03	1,22	-0,22	0,81	2,05
	(0,001)	(1,5)	(2,6)	(-0,8)	(0,2)	(0,6)	(3,1)	(-0,8)		
	9,86	-0,31	-0,05				1,15	-0,38	0,76	1,74
	(3,5)	(-0,6)	(-0,1)				(4,5)	(-1,4)		
KU	16,78	-0,26	0,08	0,15	-0,82		0,95	-0,63	0,85	2,02
	(10,5)	(-0,5)	(0,1)	(0,1)	(-0,7)		(3,9)	(-2,6)		
	12,57	-0,18	0,36	-1,06	0,65	0,14	0,95	-0,50	0,96	2,61
	(6,0)	(-0,6)	(1,1)	(-1,4)	(0,8)	(4,8)	(6,6)	(-3,5)		
	3,32	1,18	0,60				1,20	-0,58	0,80	1,68
	(1,9)	(2,4)	(1,2)				(5,7)	(-2,7)		
SE	7,77	0,98	0,41	-0,12	-0,21		1,01	-0,71	0,88	2,03
	(3,4)	(2,0)	(0,8)	(-0,2)	(-0,3)		(5,1)	(-3,7)		
	7,54	0,72	0,24	-0,28	0,03	0,08	1,01	-0,58	0,91	2,43
	(3,4)	(1,6)	(0,5)	(-0,5)	(0,1)	(2,00)	(5,8)	(-3,2)		
	1,39	-0,26	3,37				1,02	-0,34	0,66	1,54
	(0,4)	(-0,1)	(1,8)				(3,7)	(-1,3)		
ES	6,98	-0,78	2,14	0,01	-0,26		0,96	-0,63	0,77	1,63
	(2,1)	(-0,4)	(1,1)	(0,1)	(-1,8)		(4,1)	(-2,7)		
	(*) 975,14	-0,71	4,47	0,34	-0,08	0,02	1,14	-0,14	0,63	1,83
	(0,003)	(-0,2)	(2,4)	(0,6)	(-0,2)	(0,2)	(3,4)	(-0,4)		
	2,07	0,31	1,97				0,28	-0,21	0,65	1,93
	(1,8)	(0.5)	(3,0)				(1,0)	(-0,9)		
NE	4,56	0,007	1,71	-0,24	0,12		0,24	-0,08	0,68	1,97
	(1,8)	(0,01)	(2,4)	(-0,6)	(0,3)		(0,8)	(-0,3)		
	(*) 4,02	-0,08	1,82	-0,19	0,10	0,02	0,23	-0,11	0,68	1,91
	(1,4)	(-0,1)	(2,3)	(-0,4)	(0,2)	(0,3)	(0,7)	(-0,4)		
	3,71	1,07	0,95				0,74	-0,53	0,59	2,05
	(2,7)	(1,6)	(1,5)				(3,1)	(-2,3)		
SL	(*) 4,77	0,91	0,75	0,04	-0,004		0,76	-0,51	0,59	2,07
	(1,3)	(1,0)	(0,8)	(-0,1)	(-0,01)		(2,8)	(-1,8)		
	(*) 5,08	0,83	0,67	-0,07	0,02	$5,4\cdot 10^{-3}$	0,76	-0,51	0,59	2,06
	(1,3)	(0,8)	(0,7)	(-0,1)	(0,04)	(0,1)	(2,6)	(-1,5)		
	3,03	1,15	1,55				1,18	-0,73	0,92	1,81
	(4,5)	(2,8)	(5,2)				(5,9)	(-3,7)		
NA	2,22	1,42	1,53	0,12	-0,09		1,18	-0,75	0,92	1,79
	(0,9)	(2,3)	(3,3)	(0,7)	(-0,5)		(5,3)	(-3,3)		
	(*) 459,63	1,96	1,87	0,53	0,32	0,02	1,52	-0,52	0,89	1,32
	(0,01)	(2,5)	(1,8)	(1,4)	(0,8)	(0,4)	(5,4)	(-1,9)		

(*) Convergence not achieved; 't'-statistic in parantheses

Table II: *Dependent variable:* *total gross investment rate (in %)*
Q-measure: *marginal tax-adjusted Q value* *__excluding__* *dividend taxation*
Sample period: *1965-1982*

	CONSTANT	Q_{t-1}	Q_{t-2}	WR_{t-1}	WR_{t-2}	GY	RHO_1	RHO_2	R^2	DW
	7,26	0,23	0,20				1,24	-0,61	0,80	1,75
	(7,7)	(2,1)	(1,9)				(6,0)	(-2,9)		
SF	8,80	0,21	0,23	-0,22	0,12		1,23	-0,66	0,81	1,68
	(3,2)	(1,8)	(1,7)	(-0,8)	(0,4)		(5,6)	(-2,7)		
	5,12	0,13	0,10	0,12	-0,05	0,09	1,30	-0,57	0,87	1,75
	(0,9)	(1,2)	(0,8)	(0,4)	(-0,2)	(2,2)	(7,0)	(-2,9)		
	3,56	0,11	1,86				0,90	-0,54	0,64	1,79
	(2,2)	(0,1)	(1,4)				(4,2)	(-2,5)		
SB	3,13	0,48	2,12	-0,30	0,31		0,89	-0,60	0,66	1,74
	(0,8)	(0,3)	(1,3)	(-0,8)	(0,8)		(4,0)	(-2,7)		
	(*) 698,41	0,34	2,93	0,42	-0,03	0,03	1,05	-0,05	0,42	1,86
	(0,002)	(0,1)	(1,2)	(0,6)	(-0,04)	(0,6)	(3,6)	(-0,1)		
	3,26	0,96	0,92				0,99	-0,41	0,81	1,96
	(2,5)	(2,7)	(3,0)				(3,9)	(-1,5)		
ET	3,98	0,95	0,89	-0,26	0,22		1,01	-0,40	0,81	1,98
	(1,5)	(2,5)	(2,5)	(-0,5)	(0,4)		(3,7)	(-1,4)		
	5,19	0,18	0,52	-0,01	$-1,3 \cdot 10^{-3}$	0,10	0,89	-0,34	0,90	2,70
	(1,9)	(0,4)	(1,5)	(-0,03)	(-0,003)	(3,1)	(4,8)	(-1,7)		
	5,58	0,42	0,42				1,15	-0,34	0,62	1,87
	(4,6)	(2,1)	(2,0)				(4,2)	(-1,3)		
FM	8,76	0,46	0,39	0,31	-0,60		1,26	-0,53	0,75	2,21
	(3,5)	(2,5)	(2,1)	(1,2)	(-2,5)		(4,8)	(-1,9)		
	(*) 8,39	-0,13	-0,02	0,04	-0,15	0,09	1,11	-0,65	0,81	2,14
	(5,5)	(-0,5)	(-0,1)	(0,1)	(-0,6)	(2,2)	(4,9)	(-3,0)		
	3,92	1,16	0,46				1,16	-0,40	0,89	1,81
	(3,4)	(2,9)	(1,2)				(4,6)	(-1,7)		
EM	8,95	0,91	0,18	0,39	-0,75		1,16	-0,81	0,97	1,74
	(8,8)	(4,1)	(0,9)	(2,5)	(-4,8)		(7,9)	(-5,6)		
	9,00	0,88	0,15	0,38	-0,74	$4,4 \cdot 10^{-3}$	1,16	-0,79	0,97	1,75
	(8,0)	(3,1)	(0,5)	(2,1)	(-4,0)	(0,2)	(7,2)	(-4,1)		
	4,57	0,81	0,45				1,25	-0,46	0,81	1,53
	(3,0)	(2,0)	(1,1)				(4,9)	(-1,6)		
HO	10,46	0,51	0,21	-0,22	-0,18		1,10	0,56	0,84	1,70
	(4,4)	(1,3)	(0,5)	(-0,6)	(-0,5)		(4,3)	(-2,0)		
	7,3	0,15	-0,01	-0,31	0,19	0,17	0,64	-0,19	0,97	2,09
	(5,2)	(0,7)	(-0,1)	(-1,3)	(0,8)	(6,5)	(4,9)	(-1,4)		
	5,10	0,70	1,01				1,25	-0,53	0,77	1,79
	(4,6)	(1,5)	(2,5)				(5,3)	(-2,2)		
PD	8,66	0,82	0,65	0,27	-0,55		1,14	-0,55	0,83	2,23
	(3,9)	(1,6)	(1,4)	(0,8)	(-1,5)		(4,6)	(-2,2)		
	7,27	0,52	0,62	-0,03	-0,13	0,09	1,13	-0,50	0,87	2,67
	(2,9)	(1,1)	(1,5)	(-0,1)	(-0,3)	(1,9)	(5,3)	(-2,3)		
	0,57	0,43	0,65				0,81	0,07	0,86	2,02
	(0,3)	(2,1)	(3,3)				(3,6)	(0,3)		
LG	6,11	-0,17	1,02	0,40	-0,79		-0,85	-0,31	0,93	2,27
	(21,0)	(-1,3)	(7,6)	(2,2)	(-4,5)		(-3,0)	(-1,2)		
	-5,37	0,32	0,54	0,45	$6,7 \cdot 10^{-3}$	0,04	0,79	0,13	0,91	1,98
	(-0,9)	(1,6)	(2,9)	(1,9)	(0,02)	(1,2)	(3,4)	(0,6)		
	2,08	0,51	0,43				1,13	-0,22	0,83	1,90
	(0,8)	(1,6)	(1,4)				(4,2)	(-0,8)		
TG	7,30	0,43	0,36	-0,15	-0,27		0,77	-0,39	0,89	1,81
	(8,2)	(1,2)	(1,0)	(-0,4)	(-0,7)		(2,7)	(-1,4)		
	4,99	0,32	0,39	-0,25	0,09	0,12	0,83	-0,12	0,98	1,30
	(2,9)	(2,1)	(2,5)	(-1,2)	(0,4)	(5,6)	(6,0)	(-0,9)		

(*) Convergence not achieved;'t'-statistic in parantheses

The third variable included is an accelerator variable testing whether output terms influence investment beyond the extent that they influence the Q values in the first place. As can be seen from the equations in table I and II there is no unequivocal result concerning the significance of the demand variable. Although the growth rate of output is significant in some sectors, little is gained in explaining the variation of investment on the whole. Furthermore, the autoregressive parameters remain significant, indicating that the autocorrelated error term is not due to omitted variables. When comparing both Q-measures, the Q variable *excluding* dividend taxation perform better than the Q variable *including* dividend taxation. The results indicate that firms do not take dividend tax into account when calculating the rate of return on investment outlays.[25] This leads to the cautious conclusion that firms consider taxation to be of, at best, moderate importance. Strictly speaking $\theta = 1$ presumes a kind of tax illusion in the sense that tax discrimination between dividends and retentions is not accounted for in the valuation of shares. One can argue within a Q-model where new share issues are allowed for that the alternative assumptions $\theta = 1$ and $\theta \neq 1$ imply different assumptions about the firm's equity source at the margin (Poterba, J.M., Summers, L.H. (1983)). Finally, we have calculated J-type non-nested tests (see Davidson, R. and J.G. McKinnon 1981) in order to test whether the above Q-models outperform an alternative *accelerator model*. In order to do this an explicit alternative model must be formulated. For that purpose, the capital accumulation equation can be written as:

$$\left(\frac{I}{K_t}\right) = \frac{K_t - K_{t-1}}{K_t} + \frac{D_t}{K_t} \tag{20}$$

$$\approx \ln K_t - \ln K_{t-1} + \text{const.}^{26}$$

The desired capital stock K* is given by the accelerator type relationship

$$\ln K_t^* = a + b \ln Y_t. \tag{21}$$

If we assume a partial adjustment mechanism, the actual investment rate is given by

$$\left(\frac{I}{K_t}\right) \approx \lambda(\ln K_t^* - \ln K_{t-1}). \tag{22}$$

Given these relations an expression for the investment rate can easily be derived.

$$\left(\frac{I}{K_t}\right) - \left(\frac{I}{K_{t-1}}\right) = \lambda b \Delta \ln Y - \lambda \left(\frac{I}{K_{t-1}}\right) - \text{const.} + \tag{23}$$

$$\left(\frac{I}{K_t}\right) = a + (1 - \lambda)\left(\frac{I}{K_{t-1}}\right) + \lambda b \Delta \ln \gamma_t + \varepsilon_t \tag{24}$$

Although the two models given in (19) and (24) are supposed to explain the same dependent variable, neither of them can be derived from the other by parameter restrictions. The two models being rival, the J-type non-nested test procedure can therefore be used. The results of testing the models against each other are presented in table III below. The comparison of the Q model *without* real wages as additional regressors and the accelerator model leads to no unequivocal result, i.e., no model is superior to the other. Contrary to this result the Q model *including* real factor prices rejects the rival accelerator model in nearly all sectors while the accelerator model is not able to reject the Q model. In other words: The tests suggest that the Q model *including* real wages as regressors contain more explanatory power as does the accelerator model.

V. Concluding Remarks

For an analysis of West Germany's investment experience by industry the theoretical tools of Tobin's analysis, such as the concept of the valuation ratio, prove to be very useful. The paper demonstrates how these concepts can be used to incorporate the complexities of the West German tax system into the Q ratio concept. When coupled with assumptions about the objectives of the firm, the nature of the production and the adjustment cost function, this approach provides a theoretically consistent investment demand equation. When reviewing the estimation results, Tobin's Q is clearly the primary determinant of fixed investment. On the other hand, adding demand variables generally does not seem obligatory since variations in investment can be explained quite well without it. This econometric evidence is strengthened by some recent empirical work by Connolly (1986). Employing advances in error-in-variables econometric models (see Klepper, S., Leamer, E.E., 1984) Connolly shows that the Q model is less sensitive to measurement problems than other financial investment models. If one formulates the objective of an investment study in a provocative way saying that "the investment process is far too complex for any single model to be convincing ... (and) the applied econometrician, like the theorist, soon discovers from experience that a useful model is not one that is 'true' or 'realistic' but one that is parsimonious, plausible and informative" (Feldstein, M. 1982: 289) then the emerging results from our study confirm that the Q-theory of investment satisfies Feldstein's above cited ambitious claim. Whether this conclusion is robust to further econometric work remains an open question for future research.

Table III

J-Test

TOTAL GROSS INVESTMENT RATE

Hypothesis / Industry	The Q-model (Type A or Type B)* cannot reject its rival, the accelerator model, (sig = 0,05)		The ac .lerator model** cannot reject its rival, the Q-model (Type A or Type B), (sig = 0,05)	
	Type A	Type B	Type A	Type B
WG	accepted	rejected	accepted	accepted
BG	accepted	-	rejected	-
VG	accepted	accepted	accepted	accepted
CH	rejected	rejected	accepted	accepted
KU	accepted	accepted	accepted	accepted
SE	accepted	accepted	accepted	accepted
ES	rejected	accepted	accepted	accepted
NE	rejected	rejected	accepted	accepted
SL	rejected	rejected	accepted	accepted
MA	rejected	rejected	accepted	accepted
SF	accepted	accepted	accepted	accepted
SB	rejected	rejected	accepted	accepted
FM	accepted	rejected	accepted	accepted
ET	accepted	accepted	accepted	accepted
EM	accepted	rejected	accepted	accepted
HO	accepted	accepted	rejected	accepted
PD	accepted	rejected	accepted	accepted
LG	rejected	rejected	accepted	accepted
TG	accepted	rejected	accepted	accepted

* Q-model: Type A: $(I/K)_t = a + bQ_{t-1} + cQ_{t-2} + u_t$

Type B: $(I/K)_t = a + bQ_{t-1} + cQ_{t-2} + dWR_{t-1} + eWR_{t-2} + u_t$

$u_t = rho_1 u_{t-1} + rho_2 u_{t-2} + \varepsilon_t$

** Accelerator-model: $(I/K)_t = a + b(I/K)_{t-1} + cGY_t + \varepsilon_t$

Table III

J-Test

GROSS INVESTMENT RATE OF EQUIPMENT

Hypothesis / Industry	The Q-model (Type A or Type B)* cannot reject its rival, the accelerator model, (sig = 0,05)		The accelerator model** cannot reject its rival, the Q-model (Type A or Type B), (sig = 0,05)	
	Type A	Type B	Type A	Type B
WG	accepted	rejected	accepted	accepted
BG	rejected	-	accepted	-
VG	accepted	accepted	accepted	accepted
CH	rejected	rejected	accepted	accepted
KU	accepted	accepted	rejected	accepted
SE	accepted	accepted	accepted	accepted
ES	rejected	accepted	accepted	accepted
NE	rejected	rejected	accepted	accepted
SL	rejected	rejected	accepted	accepted
MA	rejected	rejected	accepted	accepted
SF	rejected	rejected	accepted	accepted
SB	accepted	accepted	accepted	accepted
FM	accepted	accepted	accepted	accepted
ET	rejected	accepted	rejected	rejected
EM	accepted	rejected	accepted	accepted
HO	accepted	accepted	rejected	accepted
PD	accepted	accepted	accepted	accepted
LG	rejected	rejected	accepted	accepted
TG	accepted	accepted	accepted	accepted

* Q-model: Type A: $(I/K)_t = a + bQ_{t-1} + cQ_{t-2} + u_t$

Type B: $(I/K)_t = a + bQ_{t-1} + cQ_{t-2} + dWR_{t-1} + eWR_{t-2} + u_t$

$u_t = rho_1 u_{t-1} + rho_2 u_{t-2} + \epsilon_t$

** Accelerator-model: $(I/K)_t = a + b(I/K)_{t-1} + cGY_t + \epsilon_t$

1 See Jorgenson, D.W. (1963), Jorgenson, D.W., Stephanson, J.A. (1967) and Jorgenson, D.W. (1971). An application of the neoclassical investment model to West Germany is contained in König, H. (1976).

2 The q-theoretical model stems from work by William Brainard and James Tobin. See Brainard, W., Tobin, J. (1968), Tobin, J. (1969) and Tobin, J. (1982).

3 See Abel, A.B. (1983), Abel, A.B. (1984) and Abel, A.B. (1985).

4 For the rationale of θ see King, M. (1977), pp. 48-53.

5 Models with convex costs of adjustment were introduced by Lucas, R.E. (1967) and Gould, J.P. (1968).

6 We assume that $\Psi(\cdot)$ is linear-homogeneous in K and I.

7 We are requiring that the transversality condition $\lim \mu(t)\, B(t) = o$ for $t \to \infty$ holds. For an alternative model with endogenous financial policy see Chirinko, R.S. (1987).

8 Time subscrips are omitted whenever possible in order to simplify the exposition. It is assumed that $i(1-\tau) = r$.

9 Average Q is defined as $Q_a = (V(o) + B(o))/pK$.

10 The proof of the proposition is quite standard and can therefore safely be omitted.

11 We define the gross(net) investment rate, somewhat more formally, as gross(net) investment divided by the value of the year's end gross (net) capital stock (both in 1980 prices). We refer to this measure as the gross(net) investment rate. For a description of the data sources compare appendix III.

12 We have concentrated on fixed investment, setting aside the category of inventory investment. Although inventory accumulation plays an important role in cyclical movements, it does not contribute to increases in productive capacity. Furthermore it represents much less of a risk than investment in physical capital and therefore provides less opportunity to test the Q-theory of investment.

13 For a recent analysis of the variability of various demand components over the business cycle in ten OECD countries compare Hillinger, C., Reich, U., Wehner, G. (1987).

14 Exceptions are plastic and ruber products, vehicles, leather products, and paper, paper products and printing.

15 The 1960 value seems to be an outlier in all series which has been caused by the boom in the West-German stock markets in 1960/61.

16 These results are similar to those in Furstenberg, G.M. von, Malkiel, B.G., Watson, H.S. (1980) for the U.S.

17 Lags actually reported in the literature on tests of Q on investment range from eight to thirteen quarters. For the latter see Engle, F., Foley, D. (1975), pp. 643-43.

18 We analyzed the efficiency of a number of different ARMA specifications and found an AR(1) and AR(2) adjustment to be the far most helpful.

19 Investment has been scaled by K to avoid spurious correlation possible with variables trending strongly upward or downward over time.

[20] We have left out the first observation in our sample because the Q values of the year 1960 are definitely outliers.

[21] Only in a few equations the DW-statistic falls into the inconclusive region.

[22] In some equations the Q values are insignificant for constructional steel and paper, printing, and publishing, too.

[23] The detailed regression output for equipment investment is available from the authors upon request.

[24] We can therefore not reject the hypothesis of the separability of the adjustment cost function.

[25] This small impact of taxes on investment is in accordance with other recent studies based on case studies and questionaire surveys. The overall conclusion which emerge from these studies that taxation has some influence on investment only for a minority of firms. See Wittmann, F. (1986), and Alan, K.F., Stafford, L.W.T. (1985).

[26] If we assume that the depreciation rate is close to a constant.

APPENDIX I: Abbreviations of Industrial Sectors

WG: Total industry; BG: Construction; VG: Manufacturing industry; CH: Chemical industry; KU: Rubber and plastic products; SE: Extracting and processing of stones and earth; ES: Iron and steel; NE: Non-ferrous metal production; SL: Constructional steel; MA: Mechanical engineering; SF: Vehicles; SB: Shipbuilding; FM: Instrument engineering; ET: Electrical engineering; EM: Tools and finished metal goods; HO: Woodworking and woodprocessing; PD: Paper, paper products and printing; LG: Leather industry; TG: Textile industry

APPENDIX II: The Calculation of Q Values

The construction of the various time series for Q by industries is based on corporations' balance-sheet and profit and loss account data compiled and published annually by the Statistical Office (Statistisches Bundesamt (ed.): Fachserie 2, Reihe 2, Abschlüsse der Aktiengesellschaften (AG)). This statistic aggregates by branch the published figures of all nonfinancial stock companies (including limited partnerships on shares) in the FRG which - for the sake of comparability - have delivered a balance- sheet likewise in the preceding year (aggregated balance-sheet data for a broader range of West German companies are published by the Deutsche Bundesbank. For empirical work on Q employing this alternative data source see Funke, M.(1986), and Ketterer, K.-H., Vollmer, R. (1981)). Because of mergers, liquidations, etc. the absolute number of corporations within the total industry sample declined from 1354 in 1960 to 804 in 1982. For a judgement on representativeness the following figures may serve: in 1982 the sample covered 97.6 percent of existing stock companies in total industry (measured in terms of face value of equity capital). The share of sample corporations' fixed assets in fixed assets of companies of all legal structures in total industry was about 24 percent in 1982 (1960: 36 percent). Tax adjusted marginal Q in the most general variant employed is calculated from the formula

$$Q_{m,d}(t) = \frac{V(t) + \theta(t)\,(B(t) - A(t))}{\theta(t)\,(1 - k(t) - z(t))\,K_{rep}(t)} \qquad \text{(A-1)}$$

the components of which are derived as follows:

V: Market value of equity is calculated dividing the industry's dividend sum on common shares as reported in (AG) by the industry's dividend yield rate (source: Statistisches Bundesamt (ed.): Fachserie 9, Reihe 2, Aktienmärkte).

B: Value of debt is taken to be the sum of book values of long-term debt (maturity $\geq$ 4 years) and of reserve contingencies from (AG).

$\mathbf{K_{rep}}$: We define total capital stock as fixed assets (i.e. equipment (E) and structures (S)) plus inventories (IN) at replacement cost. The AG data convey historic cost figures only. In order to arrive at replacement values we multiplied the historic capital stock components by conversion factors CF_E (applied to equipment and inventories) and CF_S (applied to structures), where the conversion factors CF_i represent the relation between capital stock replacement cost to capital stock historic cost. For *total manufacturing, total industry and construction* these conversion factors are calculated directly from Statistisches Bundesamt (ed.): Fachserie 18, Reihe S. 8, Revidierte Ergebnisse 1960 bis 1984 und Fachserie 18, Reihe 1, Konten und Standardtabellen 1984. For these *three broad industry groups* K_{rep} is therefore compiled according to

$$K_{rep,i} = CF_E^i \cdot (E_i + IN_i) + CF_S^i \cdot S_i \qquad i = WG, BG, VG \tag{A-2}$$

A slightly different procedure is employed for the disaggregated manufacturing industries. Because no industry specific conversion factors CF are available for the whole sample period, the conversion factors from total manufacturing have been used as a proxy for the unknown conversion factors by industry. K_{rep} for the 16 disaggregated manufacturing industries (index i) are therefore derived according to

$$K_{rep,i} = CF_E^{man} \cdot (E_i + IN_i) + CF_S^{man} \cdot S_i \qquad i = 1, \dots 16 \tag{A-3}$$

θ: the opportunity cost of retaining earnings within the firm is defined as

$$\theta(t) = \begin{cases} (1 - m(t))/(1 - \tau_R + \tau_D), & t = (1960, \dots, 1976) \\ (1 - m(t))/(1 - \tau_R), & t = (1977, \dots, 1982). \end{cases} \tag{A-4}$$

m: the marginal personal income tax rate of the representative shareholder, is calculated as a weighted average of the marginal rates faced by recipients of capital income in various tax brackets in each year. For the years until 1976 m thus derived is taken from King, M., Fullerton, D. (eds.), (1984) (KF), pp. 172-173. For subsequent years we calculated m in the same way applying distribution weights derived from income tax statistics 1977 and 1981 (Statistisches Bundesamt (ed): Statistisches Jahrbuch für die Bundesrepublik Deutschland 1981, p. 434, and 1984, p. 449) to income tax rate structures in force in the respective years. With regard to corporate tax rates τ_R and τ_D account is taken of the supplementary surcharge (1968-1976) and the stabilization surcharge (1973/74).

k: tax free investment grant per DM invested minus investment tax per DM invested.

z: z is calculated as a weighted average of the values for equipment (z_E) and structures (z_S) from equation (A-5), using as weights the shares of equipment and building investment in the industry concerned.

$$z(t)_{i,j} = \tau(t)z'(t)_j, \; j = E,S \tag{A-5}$$

z': present value of the stream of current and future depreciation allowances (including special allowances) per DM invested in t that is

$$z_j'(t) = \sum_{s=1}^{T_j(t)} D_i(t,s) \cdot (1 + r(t))^{-s+1} \tag{A-6}$$

where D(t,s) represents the depreciation rate applicable to the purchase price of assets with age s acquired in t, **r** is the discount rate (yield of fixed interest bearing securities; source: Deutsche Bundesbank (ed.): Deutsches Geld- und Bankwesen in Zahlen 1876-1976, Frankfurt a.M. 1976, p. 307, and Monatsberichte der Deutschen Bundesbank Vol. 36 (1984), No. 1, p. 57), and **T(t)** is the asset's average tax life for equipment or structures respectively (source: KF, pp. 162-163). Taking account of the fact that the investment tax raises the depreciation base and assuming that firms switch from geometrically declining-balance to straight-line depreciation of equipment in period $n \geq T(t) + 1 - 1/d$ (i.e. when for the first time the straight-line rate applicable for the rest of the asset's tax life in the case of switching over surmounts the declining-balance rate) (A-6) for equipment becomes (suppressing time subscripts)

$$z_E'(t) = (d + h)(1 + g) + \sum_{s=2}^{n^*-1} \frac{d(1 - d)^{s-2} (1 + g - (d + h)(1 + g))}{(1 + r)^{s-1}} \tag{A-7}$$

$$+ \sum_{s=n^*}^{T} \frac{\{(1 - d)^{n^*-2} (1 + g - (d + h)(1 + g))\}/(T - n^* + 1)}{(1 + r)^{s-1}}$$

h is the rate of special depreciation allowances in the first year,
g is the rate of investment tax, and
d is the declining-balance rate,

$$d(t) = x(t)/T_E(t), \tag{A-8}$$

where x is 2 from 3/1960 until 8/1977, 2.5 until 9/1981 and 3 afterwards. Additionally, account was taken of the several temporary suspensions of declining-balance depreciation. For buildings a special form of declining-balance depreciation (which might be termed stepwise linear) has been allowed in most years after 1964 as an alternative to straight-line depreciation. Since the superiority of this method from the firm's point of view is by no means obvious a priori we calculated z_S' for both methods and assumed that firms always choose the more advantageous method.

Using (A-6) straight-line z_S' is computed with

$$D(t,s) = (1 + g(t))/T_S(t), \quad s = (1, \ldots, T_S(t)), \quad t = (1960, \ldots, 1982) \tag{A-9}$$

and stepwise linear z_S' with

$$D(t,s) = (1 + g(t)) * \begin{cases} .035 & \text{for } s = (1,\ldots,12) \\ .02 & \text{for } s = (13,\ldots,32) \\ .01 & \text{for } s = (33,\ldots,50) \end{cases} \quad t = (1965,\ldots,1973, 1977,\ldots,1980) \tag{A-10}$$

$$D(t,s) = \begin{cases} .05 & \text{for } s = (1,\ldots,8) \\ .025 & \text{for } s = (9,\ldots,14) \\ .0125 & \text{for } s = (15,\ldots,50) \end{cases} \quad t = (1981; 1982)$$

Information on g,k,h,τ, and D is compiled from: Bundesministerium der Finanzen (ed.): Finanzbericht (various issues); Sachverständigenrat zur Begutachtung der gesamtwirtschaftlichen Entwicklung: Jahresgutachten (various issues); Heinzel, K. (1962); Gumpel, H.J. (1969); Gerstenberger, W. et al. (1984) (IFO); KF; Leibfritz, W., Meurer, C. (1985).

τ: in (A-4), the tax rate on corporate profits, is composed of the corporation tax rate on retained earnings and the local profits tax ('Gewerbeertragsteuer') rate. Taking account of the deductibility of local tax liability from the corporation tax base we have

$$\tau(t) = \tau_R(t)\,(1 - \tau_L(t)) + \tau_L(t). \tag{A-11}$$

τ_L: is the effective local profits tax rate for which - since this tax in fact diminishes its own tax base -

$$\tau_L(t) = .05H(t)/(1 + .05H(t)) \tag{A-12}$$

holds, where **H** is a multiplier set by the municipal tax authorities. In calculating τ_L from (A-11) each year's average H over all municipalities (source: IFO, Anhang 80) is used.

A: The present value of future tax savings due to depreciation allowances on pre-existing capital stock is the sum of A_E and A_S from equation (A-13).

$$A_i(t) = r(t)^* \sum_{s=t}^{t+\hat{T}_i} \{ \sum_{V=\hat{T}_i}^{-1} [D_i(t+v,s-t-v)I_i(t+v)]^*(1+r(t))^{t-s} \} \tag{A-13}$$

$i = (S,E)$,

$\hat{T}$: Maximum tax lifetime ($\hat{T}_E = 14$; $\hat{T}_S = 50$).

Gross investment I is taken from (AG). Since in general the number of firms in the sample declines over time it is implicitly assumed that despite of this the future tax savings due to past investment made by vanishing firms remain within the industry. Lacking investment data for the years prior to 1953 we had to assume.

$$D(x,y) = 0 \text{ for } x = 1953 \tag{A-14}$$

in applying (A-13).

Finally, a listing of the various Q definitions used may serve readers' convenience:

- Average Q

$$Q_a = \frac{V + B}{K_{rep}}$$

- Marginal tax-adjusted q ignoring different tax treatment of dividends vis-â-vis retentions ($\theta = 1$)

$$Q_m = \frac{V + B - A}{(1 - k - z)K_{rep}}$$

- Marginal tax-adjusted Q taking account of different tax treatment of dividends ($\theta \neq 1$)

$$Q_{m,d} = \frac{V + \theta\,(B - A)}{\theta(1 - k - z)K_{rep}}$$

APPENDIX III: Variables and Data Sources

I: Gross fixed investment in 1980 prices. Source: Statistisches Bundesamt (ed.): Fachserie 18, Reihe S. 8, Revidierte Ergebnisse 1960 bis 1984 (VGR); **K**: Gross capital stock in 1980 prices. Source: For total industry, total manufacturing and construction the figures are taken from VGR. Because disaggregated capital stock figures from the Statistical Office in Wiesbaden are not available for the years 1960-1969 we have linked the VGR figures from 1970 to 1982 with disaggregated capital stock estimates of the RWI-Institute in Essen (Schmidt, J. 1982, "Das Anlagevermögen in der Bundesrepublik Deutschland", RWI-Papier No. 17); **wr**: Real wage; wr = (w/p); **w**: Average hourly nominal wage of blue-collar worker. Source: 1960-1975: Statistisches Bundesamt (ed.), Fachserie D, Reihe 1, Beschäftigung und Umsatz, Brennstoff- und Energieversorgung der Industrie; 1976-1982: Statistisches Bundesamt (ed.), Fachserie 4, Reihe 4.1, Beschäftigung, Umsatz und Energieversorgung der Unternehmen und Betriebe im Bergbau und im Verarbeitenden Gewerbe (ohne Handwerk); **p**: Implicit price deflator gross value added. Source: VGR; **Y**: Gross value added in 1980 prices. Source: VGR.

REFERENCES

ABEL, A.B. (1983): "Optimal Investment under Uncertainty", in: *American Economic Review*, 73, 228-233.

ABEL, A.B. (1984): "The Effects of Uncertainty on Investment and the Expected Long-Run Capital Stock", in: *Journal of Economic Dynamics and Control*, 7, 39-53.

ABEL, A.B. (1985): "A Stochastic Model of Investment, Marginal q and the Market Value of the Firm", in: *International Economic Review*, 26, 305-322.

ALAM, K.F., AND L.W.T. STAFFORD (1985): "Tax Incentives and Investment Policy: A Survey Report on the United Kingdom Manufacturing Industry", in: *Managerial and Decision Economics*, 6, 27-32.

BRAINARD, W., AND J. TOBIN (1968): "Pitfalls in Financial Model Building", in: *American Economic Review, Papers and Proceedings*, 58, 99-122.

CHIRINKO, R.S. (1987): "Tobin's Q and Financial Policy", in: *Journal of Monetary Economics*, 19, 69-87.

CONNOLLY, R.A. (1986): "Assessing the Importance of Measurement Error in Capital Investment Models", in: *Managerial and Decision Economics*, 7, 177-185.

DAVIDSON, R., AND J.G. McKINNON (1981): "Several Tests of Model Specification in the Presence of Alternative Hypotheses", *Econometrica,* 49, 781-793.

ENGLE, R.F., AND D. FOLEY (1975): "An Asset Price Model of Aggregate Investment", in: *International Economic Review,* 16, 625-647.

FELDSTEIN, M. (1982): "Inflation, Tax Sales and Investment: Some Econometric Evidence", in: *Econometrica,* 50, 825-862.

FISCHER, S., AND R.C. MERTON (1984): "Macroeconomics and Finance: The Role of the Stock Market", in: *Carnegie-Rochester Conference Series on Public Policy,* 21, 57-108.

FUNKE, M. (1986): "Tobin's q and Industry Investment in West Germany", IIM/LMP discussion paper No. 86-13.

FURSTENBERG, G.M. von, B.G. MALKIEL, AND H.S. WATSON (1980): "The Distribution of Investment between Industries: A Microeconomic Application of the 'q' Ratio", in: Furstenberg, G.M. von (1980) *"Capital, Efficiency and Growth",* Cambridge 1980, 395-460.

GERSTENBERGER, W. et al. (1984): "Staatliche Interventionen". (*Ifo Studien zur Strukturforschung* 4), München.

GOULD, J.P. (1968): "Adjustment Costs in the Theory of Investment of the Firm", in: *Review of Economic Studies,* 35, 47-55.

GUMPEL, H.J. (1969): "Taxation in the Federal Republic of Germany", 2nd Ed. New York/Chicago/Washington.

HAYASHI, F. (1982): "Tobin's Marginal q and Average q: A Neoclassical Interpretation", in: *Econometrica,* 50, 213-224.

HEINZEL, K. (1962): "Die einkommen- und körperschaftsteuerlichen Sondermaßnahmen zur Förderung der betriebsinternen Kapitalbildung in der Bundesrepublik Deutschland seit der Währungsreform. Ein fiskalpolitischer Beitrag zur deutschen Steuerpolitik", Diss., Frankfurt am Main.

HILLINGER, C., U. REICH, AND G. WEHNER (1987): "Konjunkturschwankungen der Aggregate der VGR in 10 OECD Ländern", in: *Jahrbücher für Nationalökonomie und Statistik,* 203, 217-224.

JORGENSON, D.W. (1963): "Capital Theory and Investment Behavior", in: *American Economic Review,* 52, 247-260.

JORGENSON, D.W. (1971): "Econometric Studies of Investment Behavior: A Survey", in: *Journal of Economic Literature,* 9, 1111-1147.

JORGENSON, D.W., AND J.A. STEPHANSON (1967): "Investment Behavior in U.S. Manufacturing 1947-1960", in: *Econometrica,* 35, 169-220.

KETTERER, K.-H., AND R. VOLLMER (1981): "Bestimmungsgründe des Realzinsniveaus. Eine Untersuchung für die Bundesrepublik Deutschland", in: Ehrlicher, W. (ed.) *"Geldpolitik, Zins und Staatsverschuldung",* Berlin, 55-102.

KING, M.A. (1977): "Public Policy and the Corporation", London.

KING, M.A., AND D. FULLERTON (eds.) (1984): "The Taxation of Income from Capital. A Comparative Study of the United States, the United Kingdom, Sweden, and West Germany", Chicago/London .

KLEPPER, S., AND E.E. LEAMER (1984): "Consistent Sets of Estimators for Regressions with Errors in all Variables", in: *Econometrica*, 52, 163-183.

KÖNIG, H. (1976): "Neoklassische Investitionstheorie und Investorenverhalten in der Bundesrepublik Deutschland", in: *Jahrbuch für Nationalökonomie und Statistik*, 190, 316-348.

LEIBFRITZ, W., AND C. MEURER (1985): "Steuerliche Investitionsförderung im internationalen Vergleich", (*Schriftenreihe des Ifo-Instituts für Wirtschaftsforschung* Nr. 117), Berlin/München.

LUCAS, R.E. (1967): "Adjustment Costs and the Theory of Supply", in: *Journal of Political Economy*, 75, 321-334.

POTERBA, J.M., AND L.H. SUMMERS (1983): "Dividend Taxation, Corporate Investment, and 'Q'", in: *Journal of Public Economics*, 22, 135-167.

SUMMERS, L.H. (1981): "Taxation and Corporate Investment: A Q-Theory Approach", in: *Brooking Papers on Economic Activity*, No. 1/1981, 67-140.

TOBIN, J. (1969): "A General Equilibrium Approach to Monetary Theory", in: *Journal of Money, Credit, and Banking*, 1, 15-29.

TOBIN, J. (1982): "Money and Finance in the Macroeconomic Process", in: *Journal of Money, Credit, and Banking*, 14, 171-204.

WITTMANN, F. (1986): "Der Einfluß der Steuern auf die Investitionsentscheidungen der Unternehmen. Eine empirische Analyse", Frankfurt/New York.

YOSHIKAWA, H. (1980): "On the 'q'-Theory of Investment", in: *American Economic Review*, 70, 739-743.

Economic Policy, Profits, and Investment Behavior in West Germany

Peter Trapp

Institut für Weltwirtschaft, 2300 Kiel, West Germany

1. The Fallacy of Macroeconomic Profit Indicators

In recent years, the conditions for sustained economic growth seem to have improved noticeably in West Germany. Since the recession at the beginning of the eighties, overall profits have recovered dramatically. From 1982 to 1986, income from entrepreneurial activity and property rose by almost 50 p.c. The share of wages in national income dropped by about 5 percentage points. According to standard economic reasoning the redistribution from wages to profits should have ignited a real investment boom leading to faster growth and higher employment. The more so, since there were no bottlenecks forcing economic policy to restrain the upswing. On the contrary, the economy had returned to price level stability and interest rates had fallen to historically low levels. The public sector deficit had been narrowed, and the current account was running a huge surplus, indicating that there was ample room for expansion.

However, in spite of the favorable environment the upswing remained weak and uneven. From 1982 to 1986, real GNP rose at an average annual rate of only 2.4 p.c. In the four years following the recession of 1975, the increase was in the order of 4 p.c., and in the upswing after 1967, GNP growth amounted even to some 5.5 p.c. In particular, investment spending was disappointingly sluggish in the first half of the eighties and has hampered the increase of employment (Figure 1). Whereas the average annual increase of fixed investment of 4.8 p.c. (7.2 p.c.) was noticeably higher than the expansion of overall output in the four-year period after 1975 (1967), it has been appreciably lower than GNP growth since 1982 (1.7 p.c.). In view of this development some economists have argued that the role of profits has been exaggerated and that a "two-handed approach" providing for higher profits and more demand is necessary to achieve a faster growth of investment and higher employment [1].

Corporate Investment and Employment

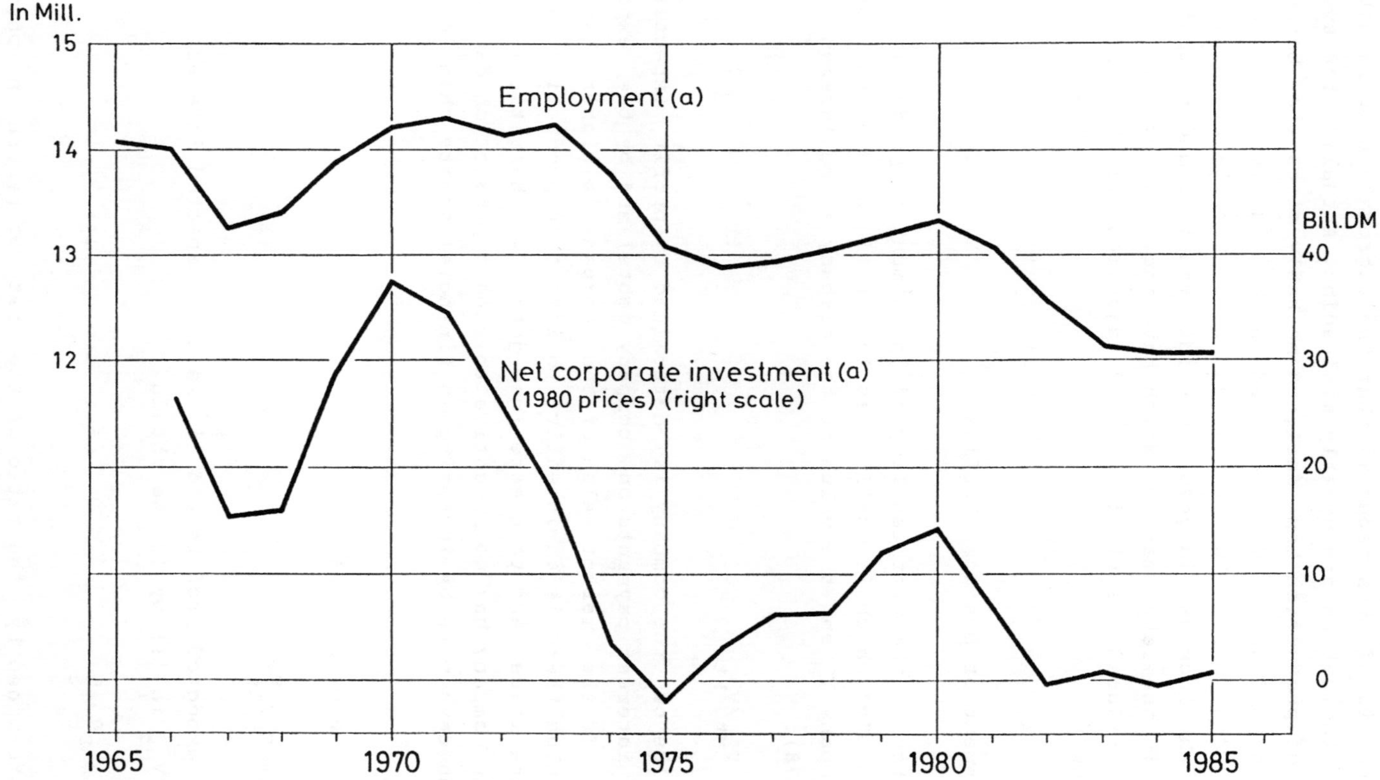

(a) Manufacturing, construction, trade.

Source: Federal Statistical Office.

The main hypothesis of this paper is that the aggregate numbers usually used as profit indicators are greatly misleading. Actually, the average profitability of production has declined markedly since the beginning of the seventies and has hardly recovered in the eighties. It is argued that the main reason for the deterioration of profits stemming from production is the increase in government interventions and regulations, in particular in the labor market, which have increased overall production costs and reduced the flexibility of adjustment to changes in demand conditions.

2. The development of Business Profits

A good indicator of the profitability of investment in productive capital is Tobin's "q" which is defined as the ratio between the market value of a company V_M and the value of the company at replacement cost V_R (Tobin, 1978).

$$q_t = V_{Mt}/V_{Rt} \qquad (1)$$

The market value of the company is determined by profit expectations (adjusted for interest payments on company debts) and by the expected interest rate. If the market value of the company is higher than its replacement value there is an incentive to invest in productive capital to enlarge capacities and to create new jobs. If the ratio is lower than unity the company has no incentive to expand production capacities because it can earn more by investing in financial assets than in real assets.

The market value of a company is

$$V_{Mt} = E^e_t/i^e_t \qquad (2)$$

where E^e are expected profits and i^e is the expected interest rate. Substituting V_{Mt} in (1) by (2) we obtain

$$q_t = r^e_t/i^e_t \qquad (3)$$

Thus, q is determined by the ratio of the rate of return on physical capital r^e_t and the rate of return on financial assets.
The problem in calculating q is to determine E^e and i^e because there

are no observations of expected magnitudes. Since we are not interested in the profit expectations of an individual company or of a certain project but in aggregate corporate profit expectations, we need an indicator of overall production profits. For Germany, the data on profit income in the National Income Accounts are very sketchy and do not allow a detailed analysis of profits actually earned by production activity. Instead we have used the balance sheet data based on annual reports of German corporations [5,6]. The balance sheet statistics cover those economic areas "in which the bill of exchange is significant as a means of finance". They basically comprise the manufacturing and the construction industry as well as trade; these sectors account for about 55 p.c. of the gross value added of all companies. The data of individual corporations are collected by the Deutsche Bundesbank and are aggregated for the sectors as a whole. Thus, the numbers give a rather comprehensive picture of the development in the sectors concerned. A comparison of the sectoral capital stock data of the National Income Accounts with the balance sheet data shows a great deal of conformity between the series. The annual surplus of the balance sheet statistics can be considered as a reasonable proxy for the expected real profit because, in general, German companies try to provide a steady stream of dividend payments to their shareholders and are keen to smoothen short run profit fluctuations in their annual reports.

It has been argued that accounting profits have a downward bias because the depreciation based on company accounts is higher than the "true" economic depreciation [4]. Notwithstanding the difficulties to determine economically justified depreciations, the argument is only valid if permissible depreciations exceed the book value of the asset. This possibility is given only for a few investment goods (e.g. ships). In all other cases, aggregate depreciation contains a relatively high depreciation on new capital goods and a relatively low (or no) depreciation on older capital goods. Thus, the argument refers more to the time pattern of depreciation of an individual asset than to the appropriateness of the aggregate depreciation number.

Concerning the interest-rate, the long-term rate on bonds which is less affected by short-run influences from monetary policy is assumed to be a measure for the expected interest rate. Since profits are defined in real terms, the interest rate is adjusted for inflation by subtracting the percentage change of the GNP deflator. Thus, the ratio between the market value of the companies' aggregate capital stock and its replacement costs will be measured by

$$q = r_t/i_t^r \qquad (4)$$

where r_t is the balance sheet rate of return on physical capital and i_t^r the real rate of interest in the bond market.

For a proper calculation of r_t some adjustments of the balance sheet data are necessary[1]:

1. The share of equity capital in total capital has declined from 30 p.c. in the sixties to less than 20 p.c. in the eighties, mainly due to relatively high costs of raising equity capital and to the taxation of such capital [20]. Therefore, interest payments have increased strongly and reduced the annual surplus. In order to eliminate the profit-dampening effect of the change in the financing structure of companies, interest payments have been added to the annual surplus.

2. Many companies that are included in the statistics are unincorporated firms. In these cases, the work of the owner and of family members in the business has to be taken into account in determining the surplus. Therefore, the income of self-employed and of helping family members has been deducted from the annual surplus.

3. According to accounting practices, the net capital stock in the balance sheet statistics is valued at historical purchasing prices. However, due to inflation the capital input actually required to earn the production profits in a certain period is considerably higher. Therefore, the net fixed capital stock has been converted to replacement costs by applying the price index for fixed capital used in the National Income Accounts.

4. Accordingly, depreciations have to be calculated from replacement values in order to eliminate inflation-induced profits (illusory profits).

5. Finally, the annual surplus does not only contain profits earned by production activities (ordinary profits) but also interest revenues and extraordinary revenues from the sale of assets and other activities. In order to separate the ordinary profits it is necessary to deduct interest earnings and revenues not related to production (neutral yields) from the annual surplus and to adjust ordinary profits for the costs originating from non-production activities.

TABLE I

Annual Surplus and Adjusted Gross Total Profits

	Annual Surplus	Gross total profits[1]	Adjusted gross total profits[2]
		In billions of DM	
1965-69	80.8	94.1	75.2
1970-74	93.6	122.1	93.3
1975-79	121.9	154.3	102.5
1980-84	134.5	191.8	124.9
1980	142.2	193.2	130.0
1981	125.5	190.9	125.6
1982	124.0	188.3	121.2
1983	137.5	187.9	120.7
1984	147.8	198.5	127.1
1985	163.0	215.0	142.1

[1] Annual surplus pre-tax and plus interest payments. - [2] Gross total profits adjusted for income of self-employed persons and illusory profits.

Sources: [5,6]; own calculations.

The adjustment of the annual surplus for interest payments, the income of self-employed and illusory profits results in slightly lower profit numbers but not in a significantly different development of profits (Table I). From the second half of the sixties to the first half of the eighties, the average annual surplus increased by 66.5 p.c. and the adjusted gross total profit by 66.0 p.c. However, there has been a dramatic change as to the sources of profits. Whereas ordinary profits accounted for two thirds of the total profits in the sixties, they contributed only roughly a quarter in the mid-eighties (Table II). Profits earned in production have even declined in absolute terms while neutral yields have quadrupled and are now a main source of income for companies. This process started in the early seventies and continued up to the first half of the eighties. In 1982, the ordinary profit share attained a low point, it has increased somewhat since then. As cyclical factors have contributed to the turnaround, it is too early to judge whether the increase will be permanent.

The decline of ordinary profits is not in contradiction to the hausse at the stock exchange from 1983 to 1986. Share prices reflect the expected earnings capacity of the company as a whole and not only the profitability of domestic production. In fact, while there was a lasting and marked decline of the ordinary profit rate, the gross profit rate remained at a fairly high level (Table III). This indicates that companies, on average, have successfully invested in financial assets and in existing or new production capacities abroad. Thus, low ordinary profits in domestic production should not be considered as being a problem of individual companies but rather as a problem of the economy as a whole. They reflect that the competitiveness of West Germany as a production location has declined and is now too low for many production activities. The yield ratio for domestic production (q) which was clearly above unity in the sixties and early seventies, fell rapidly and was even below unity in the first half of the eighties. Similar results are reported by [11]. This change in profitability has left its mark on the investment behavior of firms. The share of net fixed capital in total assets declined from 38 p.c. in the sixties to 28 p.c. in the mid-eighties (Table IV); the share of financial assets rose correspondingly.

Thus, given the development of ordinary profits in relation to yields earned on financial assets ,it is not surprising that net corporate investment was extremely subdued in recent years.

TABLE II

Ordinary Profits and Neutral Yields of German Corporations, 1965-1985

	Gross ordinary profits[1]		Neutral Yields[3]	
	in bill. of DM	in p.c. of total profits[2]	in bill. of DM	in p.c. of total profits[2]
1965-1969	50.0	66.5	25.2	33.5
1970-1974	51.3	55.0	42.0	45.0
1975-1979	40.2	39.2	62.3	60.8
1980-1984	36.9	29.5	88.0	70.5
1980	56.7	43.6	73.3	56.4
1981	38.9	31.0	86.7	69.0
1982	26.7	22.0	94.5	78.0
1983	32.2	26.7	88.5	73.3
1984	31.0	24.4	96.1	75.6
1985	37.4	26.3	104.7	73.7

[1]Profits originating from production activity. - [2]Adjusted total gross profits. - [3]Interest revenues and other revenues net of expenses.

Source: [5,6]; own calculations.

TABLE III

Gross Profit Rate, Ordinary Profit Rate, and Yield Coefficient

	Gross Profit Rate[1]	Ordinary Profit Rate[2]	Real Interest Rate[3]	Yield Ratio q[4]
1965-1969	21.5	11.5	4.4	2.6
1970-1974	17.3	7.3	2.2	3.3
1975-1979	16.4	4.3	3.0	1.4
1980-1984	15.8	3.1	5.1	0.6
1980	16.5	4.9	3.8	1.3
1981	15.8	3.2	6.6	0.5
1982	15.7	2.2	4.7	0.5
1983	15.3	2.6	4.8	0.6
1984	16.0	2.5	5.8	0.4
1985	17.0	3.0	4.7	0.6

[1]Annual surplus pretax and including interest expenditures, in p.c. of the productive capital at replacement costs. - [2]Gross ordinary profits in p.c. of the productive capital at replacement costs. - [3]Bond yield adjusted for the increase of the GNP deflator. - [4]Ordinary profit rate in relation to the real interest rate.

Source: [5,6]; own calculations.

TABLE IV

Corporate Investment in Fixed Capital and in Financial Assets

	Net fixed capital in p.c. of total assets	Productive capital[1] in p.c. of total assets	Financial assets in p.c. of total assets
1965–1969	38.3	64.8	35.2
1970–1974	35.6	61.1	38.9
1975–1979	32.6	60.1	39.9
1980–1984	29.5	57.9	42.1
1980	30.0	59.0	41.0
1981	29.8	58.5	41.5
1982	29.8	58.1	41.9
1983	29.6	57.5	42.5
1984	28.5	56.3	43.7
1985	28.2	55.7	44.3

[1]Fixed capital, inventories, currency and demand deposits.

Source: [5,6]; own calculations.

3. Why Profits Collapsed

The decline of profits can be traced back to the change in the economic policy regime that has taken place in the late sixties and has been worked out during the seventies. The political change in 1982 has not led to a reversal of the legal and institutional set up that had been developed during the last two decades.

Since the late sixties the government has moved towards more interventionist macroeconomic policies because there was widespread belief that a more active macroeconomic policy was necessary to maintain a high rate of growth and full employment; the market economy was considered to be inherently unstable. In addition, the federal government assumed responsibilities for industrial and regional developments and introduced far-reaching regulations on job safety, environmental protection, etc. Furthermore, social reforms were implemented in order to achieve "social justice" and "social peace" which were regarded as a precondition for political stability and for sustained economic growth.

An outstanding element of the new policy regime was the unconditional "full employment guaranty" which the Federal Government declared at a time when the unemployment rate was below one percent. By taking on responsibility for full employment the government increased collective moral hazard of both employers' associations and unions. They implicitly expected demand management as well as structural and social policies to take care of economic growth and full employment. By implementing a great number of programs to support growth and employment since the mid-seventies, by increasing subsidies, and by expanding labor market policies and welfare benefits, the government validated this expectation. However, the policies failed to enhance economic growth and employment. Governmental interference enabled unions and employers to pursue collectivistic wage policies that ignored the need for lower real wage increases and for greater instead of smaller wage differentials in the economy.

These policies - some of which will be outlined below - have directly or indirectly raised production costs and resulted in a pronounced decline of the profitability of domestic production.

3.1 Subsidies

Despite of its public condemnation of subsidies, German governments have given substantial support to ailing as well as to would-be sun-rise industries. According to calculations by the Institut für Weltwirtschaft based on a comprehensive definition of subsidies (financial assistance plus tax reductions), such payments to different economic sectors of the German economy have more than doubled from 1973 to 1984 [13]. Overall subsidies have risen at an average annual rate of about 7 p.c., compared to a growth rate of nominal GNP of about 6 p.c., and amounted to 6.8 p.c. of GNP in 1984 (Table V). The traditional

TABLE V

Development of Subsidies, 1973-1985

Source	1973	1980	1984
Subsidy Report of the Federal Government Billions of DM(a)	41.7	65.3	74.5
Institut für Weltwirtschaft(b) Billions of DM	56.9	102.7	120.0
Subsidies per employee in DM	2,430	4,610	5,890
in p.c. of GNP	6.2	6.9	6.8

(a) Financial assistance plus tax relief. - (c) Subsidies per person employed in the sectors subsidized. - (b) According to the definition established by the US Congress (U.S. Congress, 1972: 18).

Source: Jüttemeier [13].

recipients of subsidies were agriculture, coal mining, the federal railway system, and residential construction. However, in the course of the seventies subsidies to many other branches, e.g. shipyards, iron and steel industry, airbus industry, expanded rapidly. Public assistance to specific sectors has contributed to real wage rigidity. In particular, the response of wage determination to competitiveness problems has been very weak or even nonexistent because both employees and unions relied on the government to intervene. Wage settlements for the ailing shipbuilding industry for example are the same as for the flourishing investment goods industries. Since non-competitive industries are often regionally concentrated, employment problems in those areas are aggravated. At the same time, there is no incentive for other companies to move into those regions because labor costs are kept at an artificially high level. The state-owned shipyard Howaldtswerke-Deutsche Werft AG (HDW) provides an extreme case of how public assistance can distort competitiveness and deter investment. While hourly labor costs (including overheads) at medium-scale shipyards which were privately owned and which received a limited amount of subsidies were in the order of 45-50 DM in 1984, wage costs at HDW were as high as 75-80 DM. Because of the high wage level HDW could attract qualified personnel more easily than its competitors. At the same time HDW was able to underbid its competitors, not because productivity at HDW was much higher but because the shipyard could rely on the tax payer to settle any losses [13, p. 21].

3.2 Social Policy

In the late sixties the welfare system in West Germany was already rather comprehensive. Nevertheless, at that time the government decided that it was time to take more care of those who had not fully participated in the economic miracle to provide protection against individual risks and to give more support to families. Since it is impossible to give a complete description of all measures that have been adopted since the late sixties, only some examples will be mentioned. In aggregate, public social benefits to private households rose from 18 p.c. of GNP in 1965 to some 27 p.c. in the period from 1975 to date. If the benefits that are paid by companies are taken into account, transfers to private households rose from 25 p.c. of GNP in 1965 to about 34 p.c. in the second half of the seventies. These numbers may give an idea of the profound impact of social policy on the economy.

Social policies, which have to be financed by taxes or public debt, impose a considerable cost burden on the economy. But at least as important as the direct costs are the indirect effects on the level and structure of wages and on incentives. Welfare payments compete with earned income. If welfare payments are relatively high there is an incentive not to work and unions will push for above-average wage increases for low income earners. In the following the development of the relation between welfare payments and earned income in West Germany will be examined for unemployment compensations and social aid.

a) Unemployment compensation

The unemployment insurance, which is compulsory for all persons in dependent employment (excluding civil servants), provides two kinds of compensations: unemployment benefits and unemployment aid. Generally, a worker, who has paid unemployment contributions for a certain period of time is entitled to unemployment benefits for up to twelve months. In 1975 unemployment benefits were raised from 63 p.c. to 68 p.c. of the net income earned in the period preceding unemployment. The benefits are tax-free and up to 1982 they were calculated on the basis of net earnings including overtime payments. Since 1982 monthly benefits have been slightly reduced. Overtime payments are no longer taken into account and unemployed without dependent children receive only 63 p.c. of their net income. But in other respects unemployment benefits have been extended, in particular, for the elderly unemployed; this group can now claim benefits for up to 32 months.

Unemployment aid is granted when unemployment benefits expire and no other income sources are available. In 1975, unemployment aid was raised from 53 p.c. to 58 p.c. of the net income earned in the preceding period. Payments to unemployed without children were cut to 56 p.c. in 1982. There is no time limit for the payment of unemployment aid. In addition to unemployment benefits or aid, households are entitled to other social benefits, e.g. rent allowances, if the family income is below certain limits.

The annual income loss depends on the length of unemployment. For short periods of unemployment the annual family income after tax can even be higher than without unemployment because of the progressivity of the tax schedule and because the family may be entitled to additional benefits at the lower taxable income [9].

In 1978, the effective income loss for the whole calendar year resulting from an unemployment period of three months amounted to roughly 4 p.c. for an unmarried income earner with an average income, and to about 6 p.c. for a married income earner [14]. The income loss was in the order of 12 p.c. in both cases when the unemployment period was six months, and only if unemployment persisted for the whole calendar year a loss of 32 p.c. was incurred. The changes in the unemployment insurance system implemented since 1982 have not fundamentally changed the picture.

b) Social aid

Every household or person with inadequate sources of income is entitled to welfare payments (Sozialhilfe). The claim does not depend on previous social security contributions or tax payments. Social aid payments are intended to ensure that everybody is able to afford a minimum standard of living; it consists of monthly standard cash payments plus allowances for the rent and energy costs actually incurred plus irregular allowances for the purchase of big-ticket items (clothing, furniture, TV etc.) which have been approved by the social office. The benefits are tax-free. Welfare payments were raised markedly in the early seventies. From 1969 to 1974, monthly standard payments rose by 77 p.c. compared to an increase in the net income of a skilled worker of 52 p.c. From 1975 to 1985, social aid has been regularly adjusted to changes in the price level and in real wages. Changes in the calculation procedures and in the price index used for the adjustment of social aid resulted in an accelerated increase of entitlements in the mid-eighties.

Individual benefits depend on the size of the family, on the age of the children, and on actual outlays for rent, energy and big-ticket items. Table VI summarizes the typical assistance for a family of four (two

TABLE VI

Social Aid and Net Earnings, 1982-1986

	1982	1983	1984	1985	1986
			In D-Mark		
Standard allowances of social aid					
- Father	338	342	351	371	389
- Mother	270	274	281	297	311
- 1. child, 14 years	254	257	263	278	292
- 2. child, 8 years	220	222	228	241	253
- Total	1,082	1,095	1,123	1,187	1,245
Reimbursement for rent and energy	427	457	484	515	500
Total social aid payments	1,509	1,552	1,607	1,702	1,745
In p.c. of net monthly earnings per employed incl. family allowances	73.7	74.2	75.7	79.0	77.6

Source: Boss [2, p. 50].

adults, two children). From 1982 to 1986 the ratio of social aid (excluding irregular allowances) to the average net income of an employee has risen from 73,7 to almost 78 p.c.[2] The difference between social aid and the net income that the recipients could earn in the labor market would probably be considerably smaller than the average income because the professional skills and abilities of those receiving social aid are usually below average.

The difference between the compensations and the net income that could be earned in the labor market generally gets smaller for long-term unemployed. Therefore, the incentives to take on a job do not necessarily increase when the unemployed is no longer eligible for unemployment benefits and can only claim unemployment aid. The increase of unemployment compensation is likely to have produced more voluntary unemployment and contributed to the sharp increase of social spending. Thus, unemployment compensation and social aid have reduced the supply of labor making it easier for the trade unions to enforce higher wage increases in general and for low wage earners in particular. However, as the marginal product of less qualified workers did not rise correspondingly, unemployment for this group has increased sharply contributing to the expansion of social expenditures.

Other measures that contributed to the increase of social costs are the introduction of early retirement at only slightly reduced pension benefits (1973), the extension of maternity leave, and the adoption of more generous sick leave rules in 1970.

3.3 Labor Market Policy[3]

The readiness of the government to assume the responsibility for tackling or relieving employment problems in combination with the monopolistic features of the wage bargaining system in West Germany led to a pronounced real wage rigidity. Because wage settlements were binding for all member companies of employers' associations, adjustment to declining sales and prices was concentrated on employment. Where outsiders were able to endanger the bilateral cartel between unions and employers' associations, the government stepped in to provide protection against underbidding of insiders' wages and declared contractual wages and working conditions compulsory even for outsiders.

In addition to contractual wages there are legally enforced employment costs[4], additional benefits agreed on in collective bargaining contracts[5], and fringe benefits offered voluntarily by firms[6]. These fringe benefits can change their voluntary character and become compulsory because, if granted for a certain period of time employees obtain a title to these benefits by "law of habit" (Gewohnheitsrecht).

Compared to the remuneration for the time effectively worked, the share of fringe benefits rose from less than one fifth in 1966 to more than one third in 1984. The contracted and voluntary element almost doubled from 24 p.c. to 47 p.c.[7] (Table VII). The total expenditures of the firms for these social purposes were more than six times higher in 1984 than in 1966, while in the same period the remuneration for the work effort has only tripled.

TABLE VII

Structure of Employment Costs in Manufacturing, 966-1984

	1966	1969	1975	1984
(1) Pay for effective work (DM)	9.230	11.208	19.033	30.837
(2) Indirect employment costs (DM)	4.002	5.181	12.907	25.163
in p.c. of (1)	43.4	46.2	62.8	81.6
(3) Indirect employment costs fixed by law in p.c. of (1)	19.4	21.3	30.7	34.7
(4) Collective bargaining agreements and voluntary fringe benefits of firms in p.c. of (1)	24.0	24.8	32.1	46.9

Source: [16].

Direct and indirect wage costs increase with the size of the firms's labor force (number of members of the work council, social plan requirements, information requirements, adviser on work security). Therefore, firms are reluctant to hire additional labor when demand is rising. Thus, even in upswing periods, unemployment does not fall noticeably keeping public social outlays at a high level. Furthermore, the dismissal protection act and its interpretation by the labor courts has greatly reduced the flexibility of firms to adjust employment to changes in capacity utilization. Any dismissal has to be in conformity with some social criteria. If a dismissal is found socially unfair, it is declared ineffective. Then employment continues or - if this is not feasible - severance payments have to be agreed on. These payments can amount to an annual income. Employees cannot be dismissed after the sale of a firm. This applies even when the firm is sold after having gone bankrupt[5]. In its decisions, the Federal Labor Court has explicitly put the protection of the individual job over the protection of the firm [12].

In case of any substantial changes within the firm the company is obliged to put up a social plan which provides compensations for all employees affected by the change. The legal basis is the shop rules act of 1972 that applies to firms with more than 20 employees.

Average compensation payments in social plans in dismissal cases are estimated to have amounted to roughly D-Mark 12.500 between 1975 and 1980 [19].[9] Official statistics on aggregated dismissal compensation payments (including those resulting from social plans) are only available for manufacturing. According to these data, dismissal compensation payments in manufacturing adjusted for wage increases were three times higher in 1972 than in 1966 and about ten times higher in 1982 than in 1972.

4. The Increase of the Tax Burden

Due to the increase in government activities the share of public expenditure has risen from one third to almost half of GNP over the past twenty years. In order to finance the strong expansion of public expenditures taxes and social security contributions have been raised since the mid-sixties. Public sector revenues increased from some 34 p.c. of GNP to 42.5 p.c. in the early eighties. In particular, direct taxation has risen. The average tax burden of an unmarried skilled worker in the manufacturing sector increased from 31 p.c. in 1965 to about 48 p.c. in 1987. Marginal income taxation has increased in recent years and was in the order of 64 p.c. in 1987 (Table VIII). Company taxes (corporate tax, trade tax, property tax), too, have been raised over the past 20 years. Taxation of corporate profits amounted to almost 71 p.c. in 1986 [10].

The increase in employment costs and other production costs as well as the distorted incentives to work and to invest resulted in severe economic costs in terms of output and employment foregone. These overall economic costs are mirrored in the depressed yield on fixed productive assets. Firms switched from investments in productive capacities to financial investments and to direct investments. The consequence was a sustained slow-down of the growth rate of the domestic productive potential of the economy.

Conclusion

Under such circumstances, more demand does not necessarily translate into more investment, higher growth rates, and employment [18]. For a sustained increase in productive investment profit expectations have to be improved by reducing government interventions and regulations [8]. In addition, more flexibility has to be enforced in wage policy by decreasing moral hazard in social policy and in wage negotiations and by allowing fiercer competition in product and factor markets. Last but not least, taxes on income and profits have to be cut in order to regain competitiveness of the production location West Germany [3].

TABLE VIII

Income Taxation of Typical Households of Employees in 1982 and 1987

Type of households	Average tax rate(b) (p.c.)	Marginal tax rate(b) (p.c.)
	- 1982 -	
Single monthly earnings(a) D-Mark 3,000	45.7	60.5
Family with two children		
- 1-income-earner, monthly earnings(a) D-Mark 3,000	39.5	45.8
- 2-income-earners monthly earnings(a) each DM 3,000	44.9	59.8
	- 1987 -	
Single, monthly earnings(a) D-Mark 3,558	48.6	64.3
Family with two children		
- 1-income-earner, monthly earnings(a) D-Mark 3,558	39.3	47.1
- 2-income-earners monthly earnings(a) each DM 3,558	46.6	62.8

(a) Monthly earnings including employer's contributions to social security. - (b) Income tax and social security contributions. Contributions in 1982 (1987): Old age and disability insurance 19.2 (18.7) p.c.; health insurance 12.1 (12.5) p.c.; unemployment insurance 4.0 (4.3) p.c.

Source: [3]

Footnotes

1 For a detailed description of the adjustments see [7].

2 The average income also includes the income of part-time workers. It is, therefore, lower than the average income of a full-time worker. On the other hand, actual social aid benefits are higher than in the table because irregular allowances are not included.

3 This chapter is based on the evidence presented [16] and on [15].

4 Employer's contributions to social security, paid sick leave, paid holidays and annual leave, maternity protection, nursing pauses, supplements to maternity pay, general protection against dismissal, special protection for distinct groups of employees (works council members, workers in compulsory military service, handicapped persons, apprentices), employment protection in the case of takeovers, paid time-off for educational vacation and special vacation for youth oriented work, occupational safety and health prescriptions, social plan (compulsory severance payments), costs to comply with the plant constitution act (meetings, members of the works council), special employment protection for young workers (working time, restrictions on employment), and costs arising from complying with the vocational training act.

5 Paid vacation and sick leave over legal minimum provisions, special payments (e.g. Christmas bonus), contributions to saving schemes, bonus and extra pay, agreements on short-time working, agreements on protection against rationalization, income protection for elderly workers, and other social agreements.

6 Provisions for old-age-retirement, health service, housing allowances, household and family benefits, and other fringe benefits.

7 The voluntarily agreed fringe benefits amounted to roughly one fifth of total fringe benefits in 1984.

8 This provision was adopted in 1972 after the passing of the shop rules act.

9 For comparison, from 1975 to 1980 average net earnings of employees in the West German economy (social plan payments are tax free) amounted to DM 18,500.

R E F E R E N C E S

[1] Blanchard, O. R. Dornbusch, J. Drèze, H. Giersch, R. Layard and M. Monti: Employment and Growth in Europe: A Two-Handed Approach. Bruxelles: CEPS Papers No. 21, 1985.

[2] Boss, A.: "Finanzpolitik seit 1982: Stärker wachstumsorientiert?" Die Weltwirtschaft, 1986, Heft 1, 47-54.

[3] Boss, Alfred (1987), "Zur Steuerreform in der Bundesrepublik Deutschland." Die Weltwirtschaft, 1987, Heft 1, 47-60.

[4] Chan-Lee, J. H.: Pure Profit Rates and Tobin's q in Nine OECD Countries. Paris: OECD Working Papers No. 34, April 1986.

[5] Deutsche Bundesbank: Jahresabschlüsse der Unternehmen in der Bundesrepublik Deutschland 1965 bis 1981. Sonderdrucke der Deutschen Bundesbank Nr. 5, Frankfurt/M., 1983.

[6] Deutsche Bundesbank: "Ertragslage und Finanzierungsverhältnisse der Unternehmen im Jahr 1985". Monatsberichte der Deutschen Bundesbank 38 (1986), No. 11, 13-29.

[7] Dicke, H. and P. Trapp: "Zur Ertragskraft von öffentlichen und privaten Investitionen." Die Weltwirtschaft, 1985, Heft 1, 70-87.

[8] Donges, J. B. and K.-W. Schatz: Staatliche Interventionen in der Bundesrepublik Deutschland. Kiel: Kieler Diskussionsbeiträge Nr. 119/120.

[9] Fritzsche, B. and D. von Löffelholz: "Unbeabsichtigte Einkommensvorteile bei Arbeitslosigkeit." Mitteilungen des Rheinisch-Westfälischen Instituts für Wirtschaftsforschung, Jg. 32 (1981), 13-20.

[10] Fuest, W. and R. Kroker: "Unternehmens- und Investitionsbesteuerung im internationalen Vergleich." Institut der Deutschen Wirtschaft, IW-Trends, 1987, A1-A17.

[11] Funke, M.: "Einflüsse auf die Entwicklung der Kapitalrentabilität im verarbeitenden Gewerbe in der Bundesrepublik Deutschland". Jahrbuch für Sozialwissenschaft, 38 (1987), 188-205.

[12] Giebel, U. J.: Sozialleistungen und Unternehmensgrößenstruktur, Köln, 1985.

[13] Jüttemeier, K. H.: Subsidizing the Federal German Economy - Figures and Facts, 1973-1984. Kiel: Kiel Working Paper No. 279, 1987.

[14] OECD: The Challenge of Unemployment. A Report to Labor Ministers, Paris, 1982.

[15] Soltwedel, R.: Staatliche Interventionen am Arbeitsmarkt - eine Kritik, Kiel, 1984.

[16] Soltwedel, R. and P. Trapp: "Labor Market Barriers to More Employment: Causes for an Increase of the Natural Rate? The Case of West Germany." Kiel: Paper presented at the Kiel Conference on "Macro- and Micro Policies for More Growth and Employment", 1987, forthcoming.

[17] Tobin, J.: "Monetary Policies and the Economy: The Transmission Mechanism." The Southern Economic Journal, Vol. 44 (1987), 421-431.

[18] Trapp, P.: "West Germany's Economic Policy: What Direction?" CATO Journal, 6 (1987), 837-850.

[19] Vogt, A.: Sozialpläne in der betrieblichen Praxis. Köln, 1981.

[20] Weichert, R.: "Zur Besteuerung von Risikokapital in der Bundesrepublik Deutschland", Die Weltwirtschaft, 1986, Heft 1, 89-105.

Anticipated Tax Changes and Investment in the UK: An Empirical Evaluation of the 1984 Tax Reform

Elias Dinenis

London Business School, Sussex Place, Regent's Park, London NW1 4SA, U.K.

INTRODUCTION

Throughout the postwar period a number of fiscal incentives was introduced to lower the cost of investment and promote capital accumulation in the UK. The most generous incentive was the first year allowance (FYA) which allowed firms to expense all their investment expenditure on plant and up to 75 percent of their expenditure on buildings immediately. In the 1984 Budget, the Chancellor departing from this policy announced the phasing out of the FYA with eventual abolition in 1986. The annual writting down allowance though, was retained.At the same time the corporate tax rate was to be reduced to 35 percent by 1986 in three successive cuts.

Estimates of the effects of the new measures on investment vary considerably. Edwards(1984) shows that the cost of capital will fall during the transitional period 1984–86 and it will increase from 1986 onwards. Thus investment will increase during the transitional period and fall after 1986. Although no quantitative estimates are offered ,Edwards believes that investment expenditure will be brought forward to take advantage of the reduced cost of investment, generating an investment led slump once the reform is completed in 1986. The Treasury's own estimates on the other hand HMSO(1984), suggest that investment will increase during the transitional period and it will fall after 1986 but the changes in investment attributed to the tax reform are small, of the order of 2 percent. Finally, Sumner(1986) finds that there is substantial shifting of investment to the first year of the transional period.

The purpose of this paper is to assess the short and long run impact of the measures introduced in the 1984 Budget on private nonresidential investment in the UK. We proceed in two stages. First we specify and estimate an investment model under the assumption that firms face adjustment costs in changing their stock of capital. The presence of these costs make investment a function of Tobin's Q ratio, the value of the firm's reproducible assets ,adjusted for taxes, to their replacement cost. This is done in Sections 1 and 2. In the second stage, Section 3, we calculate the effects of the tax reform on Q and then using the estimated investment equation we calculate the effect of the tax reform on investment. The impact of the tax reform on investment then depends on two factors: (a) how the Q ratio is affected and (b) how this ratio in turnthe size of the adjustment costs. If adjustment costs are high then investment will be rather insensitive to changes in Q however large these changes may be.

1. A MODEL OF INVESTMENT WITH ADJUSTMENT COSTS

Consider a firm that produces a single output using labour, raw materials and capital K which depreciates exponentially at a rate δ. The firm is a price tmarket and output market, and incurs adjustment costs with respect to investment. The adjustment costs are assumed to be internal to the firm, in that they relate not to aupward sloping supply schedule for capital goods , but the cost of diverting the resources of the firm from the production to the expansion process. These costs increase as the size of investment or disinvestment rises and are negatively related to the size of the firm. These costs can represented by the following function

$$G = G(I,K) = \varphi(\frac{I}{2K} - \delta)I \qquad (1)$$

with the properties G_I , $G_{II} > 0$ and $G_k < 0$, $G_{kk} > 0$ .In addittion G is assummed to be linearly homogeneous in I and K and to be zero when net investment is zero.The production tecnhology of the firm is described by the real restricted cost function $\pi(w,K)$ which is assumed to possess the usual neoclassical properties ie nonincreasing and concave in the real price of variable inputs w, and nondecreasing and concave in K. Moreover we assume that the technology exhibits constant returns to scale.The real cash flow of the firm in period t is given by

$$R_t = (1-\tau_t)[\pi(w,K_t) - G(I_t,K_t)] - p^k{}_t I_t - T_t \tag{2}$$

where p^k is the real price of investment goods, τ is the corporate tax rate and T is the real wage bill defined as

$$T_t = \tau_t[\pi(w,K_t) - G(I_t,K_t)] - \Sigma_{j=-\infty}^{t} D(t,j)(1+\dot{p})^{-(t-j)} p^k{}_j I_j \tag{3}$$

D(t,j) is the depreciation allowance per pound of date j investment expenditure, set off against tax liabilities in period t, and $\dot{p}$ is the rate of inflation. We ignore personal taxes and the tax system is described by D and τ. To simplify the analysis we assume that depreciation for tax purposes is calculated on a declining balance basis with special provisions for the first year.

$D(t,t) = b + d(1-b)$

$D(s,t) = d(1-b)(1-d)^{s-t}$

here d is the writting down allowance and b is the first year allowance. Writing down allowances are applied on the residue of expenditure not qualifying for first year allowance. If b=1, then investment expenditure is immediately expensed. If b=0 then there is no first year provision.

There is no uncertainty in the model and the firm's planning is done under perfect foresight. Therefore its objective is to maximise the present value of future cash flows ,discounted at the real after tax cost of capital r, which in the absence of personal taxes is a weighted average of the cost of equity and debt. In each period the value of the firm is given by :

$$V_t = \Sigma_{s=t}^{\infty} \beta^{(s-t)}(1-\tau_s)[\pi(w_s,K_s) - G(I_s,K_s) - p_s{}^k[1-\Sigma_{i=s}\tau_i(1+\dot{p})D(i,s)]I_s$$

$$+ \Sigma_{s=t}^{\infty}\beta^{(s-t)}\tau_s\Sigma_{j=-\infty}^{t-1}(1+\dot{p})^{-(s-j)}D(s,j)I_j = V_t(K,I) + A_t$$

where $\beta = 1/(1+r)$ is the discount factor and A is the present value of depreciation allowances due to past investment. That is in each period the value of the firm consists of two components. The present value of future real after tax profits V and the term A which is unaffected by the firm's current and future actions and therefore predetermined in period t.. Consequently the firm solves the following problem

Max $\overline{V}(I,K)$

subject to

$$K_t = I_t + (1-\delta)K_{t-1} \tag{6}$$

With $\lambda(t)$ the adjoint variable associated with (6), the Hamiltonian is

$$H_t = (1-\tau_t)[\pi(w_t,K_t) - G(I_t,K_t)] - (1-z_t)p_t^k I_t - \lambda_t[K_t - I_t - (1-\delta)K_{t-1}]$$

where $z_t = \tau[b + d(1-b)\Sigma_{s=t}[(1+p)\beta(1-d)]^{s-t}$ (7)

is the present value of depreciation allowances per pound of new investment.

The optimal pair of I,K should satisfy the first order conditions

$$(1-\tau_t)\varphi(\frac{I_t}{K_t} - \delta) + (1 - z_t)p_t^k = \lambda_t \tag{8}$$

$$(1-\tau_t)[\pi(w_t) + \frac{1}{2}\varphi(\frac{I}{K})_t^2] + \beta(1-\delta)\lambda_{t+1} = \lambda_t \tag{9}$$

and the transversality condition

$$\lim_{t\to\infty} \lambda_t = 0 \tag{10}$$

Let $\overline{V}(K_t^*,I_t^*)$ be the maximized value of V. Since it has been assumed that the profit is concave and the adjustment cost functions is convex in K, the Hamiltonian is concave in K. If we further assume that the Hamiltonian is also differentiable then by a well known result in optimal control theory (see e.g Seierstad and Sydsaeter (1987)), its derivative with respect to K is equal to the adjoint variable

$$\frac{\partial \overline{V}}{\partial K_t} = \lambda_t \tag{11}$$

That is the adjoint variable measures the change in the optimal value function of a unit change in K_t. The interpetation of the first order conditions equations (8) and (9) is now straightforward. Equation (9) states that the marginal value of capital equals the present value of after tax marginal profits, which are independent of the capital stock because of the linear homogeneity of the profit function. Marginal profits are positively affected by the investment ratio since a higher investment ratio will reduce adjustment costs by increasing the size of the capital stock. Equation (8) states that the firm will invest up to the point where the marginal benefit from the new investment λ equals its marginal cost, the latter including not only the price of capital goods but also all adjustment costs. The optimal investment rule is derived by solving (8)

$$\frac{I}{K} = \delta + \frac{1}{\varphi(1-\tau_t)}[\lambda_t - p_t^k(1-z_t)] = \delta + \frac{1}{\varphi} Q_t \tag{12}$$

where $Q_t = \frac{p_t^k}{(1-\tau_t)}[q_t - (1-z_t)]$

and $q = \lambda/p^k$ is Tobin's marginal q ratio defined as the marginal value of new investment to its cost.

In view of (12), equation (9) can be written as

$$(1-\tau_t)[\pi(w) + \frac{1}{2}\varphi\delta^2] + \frac{1}{2\varphi(1-\tau_t)}[\lambda_t - p_t^k(1-z_t]^2 - \delta p^k z_t + \beta\lambda_{t+1} = \lambda_t \tag{13}$$

which is a nonlinear difference equation in λ. Similarly (6) and (12) could be combined to produce a difference equation in K. The resulting difference equation together with (13) would comprise a system of nonlinear difference equations in K and λ. Phase space analysis could then be employed to investigate the effects of changes in tax policy on the steady state values of K and λ as e.g in Abel (1982). The linear homogeneity of the profit and adjustment cost functions, means, however, that the steady state optimal capital stock is not defined. An alternative way of investigating the effects of tax policy, is to treat policy changes as deviations of certain control variables from their steady state values. (see e.g. Auerbach (1986)). Linearizing (13) around the steady state values λ^*, z^* and τ^*, and solving the linearized difference equation yields

$$\lambda_t = \lambda^* + \Sigma_{s=t}^{\infty}\beta^{s-t}\alpha_s$$

where

$$\alpha_s = \delta p^k(z_s - z^*) - [\pi(w) + \frac{1}{2}\varphi\delta^2]\,\Delta\tau_s \tag{14}$$

and $\Delta\tau = \tau - \tau^*$

Also linearizing the investment equation we obtain

$$(\frac{I}{K})_t = \delta + \frac{1}{\varphi(1-\tau^*)}(\lambda_t - \lambda^*) + \frac{1}{\varphi(1-\tau^*)}(z_t - z^*) \tag{15}$$

Now we can investigate the effects of tax policy on investment.

A. <u>Anticipated Tax Rate Cut</u>

Suppose that the tax rate remains constant at a level τ up to period T and is cut permanently to τ^*. This will affect both terms in (14). From (7) we have that

$$z_s - z^* = d(1-b)\Sigma_{i=T}^{\infty}[\beta(1-d)]^{i-s}\Delta\tau \quad \text{for } s < T$$

and $$z_s - z^* = \{b + d(1-b)\Sigma_{i=0}^{\infty}\,[\beta(1-d)]^i\Delta\tau \quad \text{for } s \geq T$$

and consequently λ_t can be written as

$$\lambda_t = \lambda^* + \frac{\delta p^k d(1-b)}{1-\beta(1-d)} \frac{1-\beta^{T-t}}{1-\beta} \Delta\tau + \delta p^k [b + \frac{d(1-b)}{1-\beta(1-d)}] \frac{\beta^{T-t}}{1-\beta} \Delta\tau \tag{16}$$

$$- [\pi(w) + \frac{1}{2}\varphi\delta^2] \frac{\beta^{T-t}}{1-\beta} \Delta\tau$$

The first term in (16) represents the capital loss that holders of capital installed between period t and T−1 will experience in period T. If investment is immediately expensed (b = 1) ,this term becomes zero since there are no unclaimed tax benefits from earlier periods in period T. When the tax change takes place immediately, (T = t), again this term dissapears.Since $\tau < \tau^*$ this term will reduce the marginal valuation of capital.

The second term represents the loss in tax allowances per pound of new investment that takes place after period T. This would have a negative effect on the marginal value of capital, since lower allowances imply a lower investment rate and consequently a lower marginal product of capital. This term also becomes zero when b=1. The more distant the date of the tax change is , the small the effect of this term is on λ. The last term represents the pure profit effect and has a positive sign. It is the present value of future tax savings that would be available because of the reduction in the corporation tax rate.

The effect on investment is ambiguous. The second term in (15) is negative, whereas the first term can be either positive or negative.Thus investment in period t may increase or decrease in anticipation of a tax cut in period T.

B. Abolition of FYA in period T

In this case the marginal value of capital is given by

$$\lambda_t = \lambda^* - \tau[1 - \frac{d}{1-\beta(1-d)}] b^* \beta^{T-t} \frac{1}{1-\beta}$$

and it will reduce the marginal valuation of capital. The effect on investment is always negative. The first term in (15) is always negative. The second term is zero unless T=t in which case it will be negative reinforcing the first effect. In this case, an unanticipated abolition of the FYA will a larger effect than an anticipated one.

2. EMPIRICAL RESULTS

To evaluate the effects of the 1984 tax reform on private nonresidential investment in the U.K. we proceed in three stages. First we find an empirical counterpart of the Q term. Secondly we estimate the parameters of the investment equation by estimating equation (12) for the sample period 1953–1986 and thirdly we use the estimated investment equation to calculate the effects of the tax reform. In this section we shall describe the first and second stage.

The main problem in estimating an investment equation based on Tobin's Q theory is how to proxy the unobservable variable q. Hayashi (1982) has shown that the linear homogeneity of the adjustment cost function and of the production function is sufficient to express q as a function of observable variables. The linear homogeneity of the profit and adjustment cost functions implies that $\overline{V}(K,I)$ is also linearly homogeneous, which in turn implies that

$$\frac{\partial \overline{V}}{\partial K} = \frac{\overline{V}}{K} = \lambda$$

so that the value of the firm in every period can be written as

$$\frac{V}{p^k K} = \frac{\overline{V}}{p^k K} + \frac{A}{p^k K} = \frac{\lambda}{p^k} + \frac{A}{p^k K}$$

from which we have

$$q = \frac{\lambda}{p^k} = \frac{V}{p^k K} - \frac{A}{p^k K} \qquad (17)$$

that is the value of a unit of new capital differs from the value of a unit of existing capital only because of their different tax atributes The empiriccal counterpart of Tobin's marginal q can be calculated from (17). V is the market value of the existing capital as it is formed in the stock exchange.

Average Q

The market value of the existing capital stock was calculated by capitalizing ordinary and preferennce dividends, interest payments on long term debt and profits due abroad.

$$\frac{V}{p^k K} = adj^*(\frac{OD}{r^e} + \frac{PD}{r^p} + \frac{DB}{r^d} + \frac{PDA}{r^e})/(NK + BIN)$$

where

OD = payments of ordinary dividends

PD = payments of preference dividends

DB = payments of interest on long term debt

PDA = profits due abroad

r^e = yield on ordinary dividends

r^p = yield on preference dividend

r^d = yield on debentures

NK = beginning of period net capital stock at current replacement cost

BIN = beginning of period stock of inventories

adj = the ratio of domestic to total income

Present value of depreciation allowances (z)

Because of their different tax treatment, fixed assets are divided into plant and machinery, vehicles and buildings. Vehicles and plant and machinery depreciate for tax purposes on a declining balance method, whereas buildings on a straight line basis. In this study we treat plant and machinery and vehicles as one category. Four types of tax allowances have been in operation during the period 1950–1986. The writing down allowance z^d, the investment allowance z^i, the first year allowance z^a, and the investment grant z^g. In calculating the present value of depreciation alowances we have taken into account the announcement effects.

For plant and machinery the present value of depreciation allowances per pound of new investment is given by

$$z^p(t) = z^g(t)B(t) + \{z^i(t) + z^d(t)[1-z^g(t)] + z^a(t)\}B(t)\tau(t,t)$$

$$+ B(t)\Sigma_{i=1}^{\infty} B(t)^i \tau(t,t+i)A(t,t+i)$$

where

$$A(t,t+k) = \frac{z^d(t,t+k)}{z^d(t,t+k-1)}[1 - z^d(t+k-1)]\, A(t,t+k-1) \qquad \forall\ k \geq 1$$

and

$$A(t,t) = \frac{z^d(t,t)}{1-z^d(t,t)}\{ 1 - z^d(t,t)[1-z^g(t)] - z^a(t) \}$$

Similarly for buildings we have

$$z^b(t) = \{\tau(t,t)[z^a(t) + z^i(t)] + \Sigma_{i=0}^{T} \tau(t,t+i)z^d(t,t+i)*B(t)^i \}\, B(t)$$

and T is calculated by solving the following equation

$$z^a(t) + \Sigma_{i=0}^{T} z^d(t,t+i) = 1$$

z is a weighted average of the two z's

$$z(t) = az^p(t) + (1-a)z^b(t)$$

where a is the proportion of vehicles, plant and machinery to total investment.

<u>Depreciation allowances on past investment (A)</u>

Let PV(j,k) be the present value in period k of the remaining allowances due to investment in period j. For plant and machinery this takes the form

$$PV^p(j,k) = B(k)[\Sigma_{i=0}^{\infty}\tau(k,k+i)A(j,k+i)B(k)^i]INV_j$$

whereas for buildings the formula becomes

$$PV^b(j,k) = [\Sigma_{i=0}^{T} \tau(k,k+i)z^d(k,k+i)B(k)^i]INV_j$$

where INV stands for investment at current prices

The present value of depreciation allowances in period t due to past investment is given then by

$$PV^i(t) = \Sigma_{j=-\infty}^{t-1} PV(j,t)INV_j \text{ for } i = p,b$$

and A is defined in each period as

$$A = PV^p + PV^b$$

<u>The estimated investment equation</u>

Having derived all the empirical counterparts of the theoretical variables we can now proceed with the estimation of the parameters of the investment equation. The relationship between investment and Q in equation (12) contemporaneous and static. The existence of delivery and expectational lags though, means that the relantioship between Q and investment should be a dynamic one. Our estimation strategy consisted of estimating a general dynamic model of the form

$$A(L)(I/K) = B(L)Q_t + \Gamma(L)\epsilon_t \qquad (18)$$

where L is the lag operator and A,B and Γ are polynomials in L in order to arrive at a parsimonious representation of the relationship. The following equation with an AR(1) error structure was found to be a satisfactory parsimonious representation of (18).

$$(\frac{I}{K})_t = \underset{(20.57)}{0.03727} + \underset{(4.855)}{0.00294}\, Q_{t-1} + \underset{(15.05)}{0.8138}\, \epsilon_{t-1} \qquad (19)$$

SE = 0.00158 DW = 1.828 LM(1) = 0.616

LM(1) is the value of the Lagrange multiplier for first order serial correlation in the residuals. The implied depreciation rate is 3.7 percent and the adjustment cost parameter is 340. This value is larger than the one found by Dinenis(1985) using disaggregated data and implies large adjustment costs, making the effect of Q on investment very small.

3. THE EFFECTS OF THE 1984 TAX REFORM ON INVESTMENT

Having estimated the parameters of the investment equation, we can now evaluate the effects of the 1988 tax reform. The main changes are shown in Table 1. The first year allowance was phased out over a period of three years and the corporation tax rate was reduced from 50 percent to 35 percent over the same period. The annual writing down allowance remained unchanged at 25 percent for expenditure on plant and machinery and at 4 percent for expenditure on buildings.

TABLE 1

	FYA Plant	Buildings	Corporate Tax Rate
Expenditure incurred:			
on or before 13.3.84	100%	75%	50%
from 14.3.84 to 31.3.85	75%	50%	45%
from 1.4.85 to 31.3 86	50%	25%	40%
on or after 1.4.87	nil	nil	35%

MARGINAL q AND z

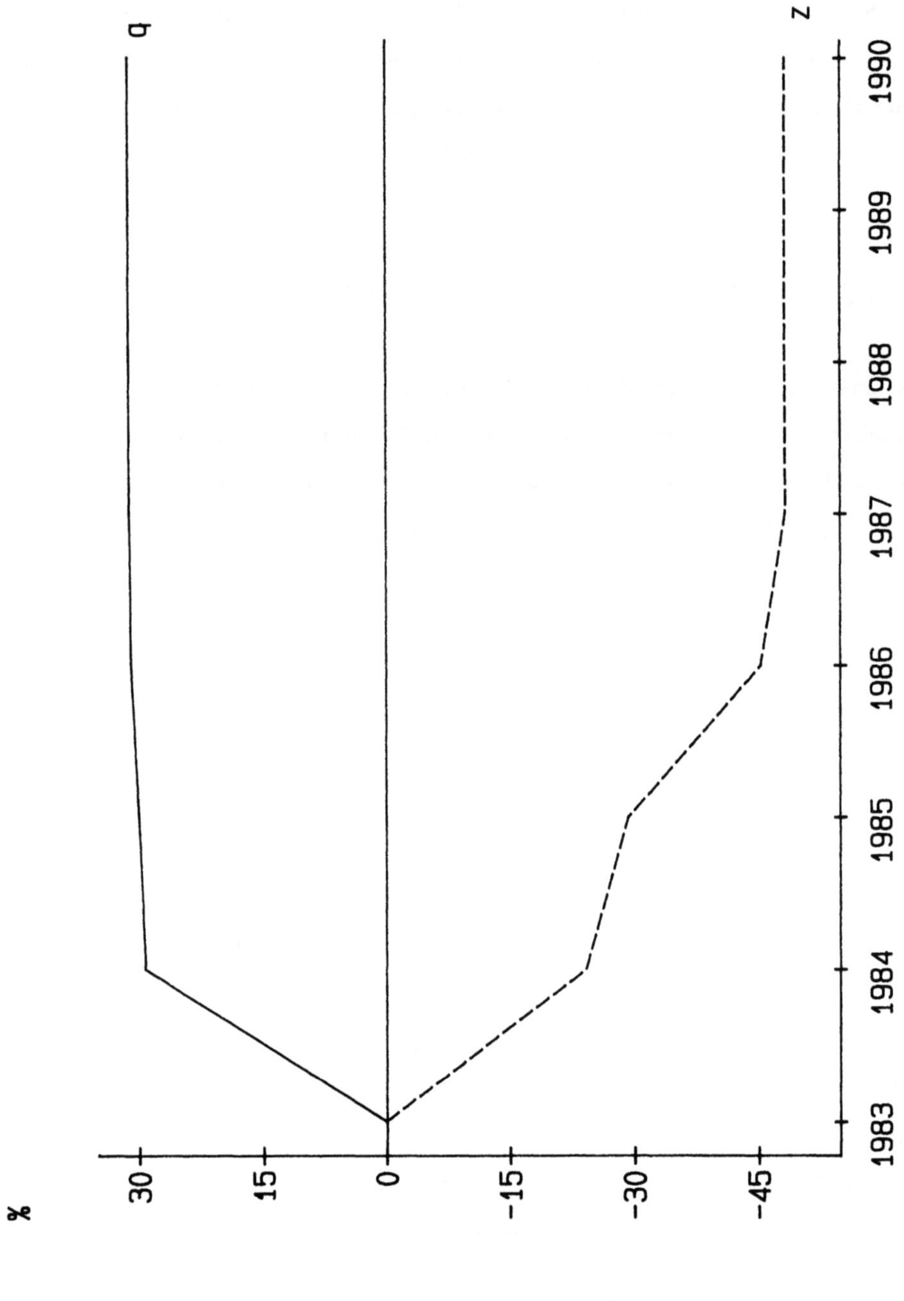

FIGURE 1

%
TAX ADJUSTED Q
22.5
15.0
7.5
0.0
-7.5
-15.0
1983
1984
1985
1986
1987
1988
1989
1990

FIGURE 2

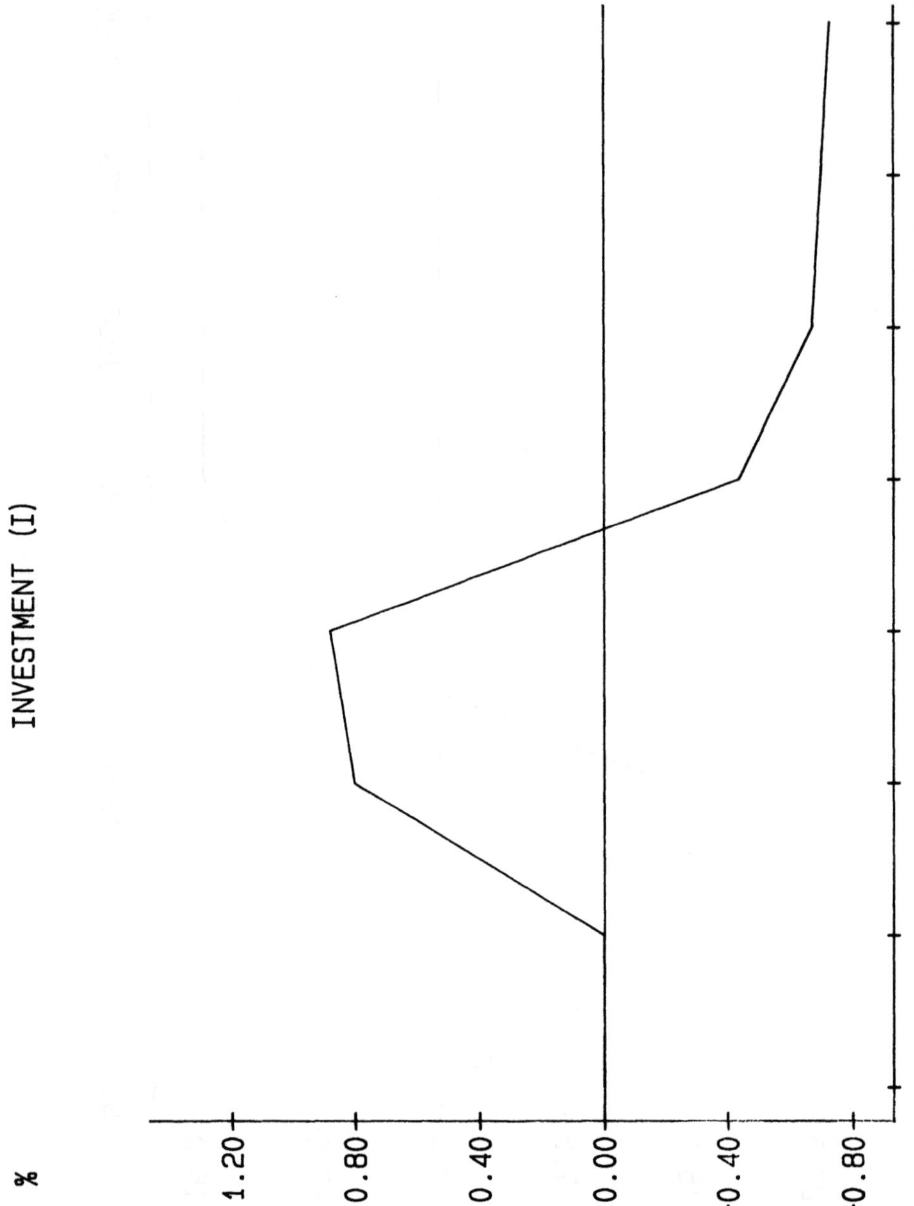
INVESTMENT (I)
%
1.20
0.80
0.40
0.00
-0.40
-0.80
1983
1984
1985
1986
1987
1988
1989
1990

FIGURE 3

As it was shown in Section 1, the cut in the corporation tax rate will increase q while the effect on Q and investment is ambigious. The abolition of the first year allowance will decrease Q and investment will fall. The net effect of the combined tax cut and the abolition of FYA can be either positive or negative. To calculate the effects of the 1984 tax reform the following model was solved over the period 1984–1990 first under the assumption that all tax parameters retain their 1983 values , and then with the parameters taking the values of the tax reform. The investment mix was assumed to remain the same.

$$(\frac{I}{K})_t = 0.00694 + 0.00294\, Q_{t-1} - 0.00239\, Q_{t-2} + 0.81376\, (\frac{I}{K})_{t-1} \quad (20)$$

$$Q_t = \frac{1}{(1-\tau_t)}[\lambda_t - p_t^k(1-z_t)]$$

$$\lambda_t = (1-\tau_t)[\pi(w_t) + \frac{1}{2}\varphi(\frac{I}{K})_t^2] + \beta(1-\delta)\lambda_{t+1} \quad (9)$$

$$z_t = az^p + (1-a)z^b$$

The presence of the lead term requires a special solution method. Perfect foresight rquires that the value of the lead term λ conforms to its realized value. To achieve this conformity, an algorithm similar to Fisher, Holly and Hughes Hallet (1986) was implemented. The algorithm consists of the following steps. First the new steady state value of λ which will prevail under the new regime was calculated. This steady state value will be the terminal value for λ, ie $\lambda_{T+1} = 1-z$, where T is the last simulation period. Secondly we guess an initial path for all the future values of λ. Thirdly, we solve the above model for Q, and I/K . Using the derived value of I/K in (9) we obtain an estimate of λ. Now we compare the initial guess with the derived variable. If they are not sufficiently close to one another, we update the path of guesses in a Jacobian fashion,

$$\lambda_t^{*k+1} = \mu\lambda_t^k + (1-\mu)\lambda_t^{*k} \qquad t = 2,3 \quad ,T$$

where k represents the iteration and μ is a parameter between 0 and 1.

The percentage deviations in q, z, Q and investment from the base run under the unchanged tax system are given in Figures 1,2 and 3. Following the anounced cut in the corporation tax rate, q will jump immediately by 29 percent and it will be in the long run higher by 31 percent. The present value of depreciation allowances (z) is reduced because of the lower corporation tax rate and the abolition of the first year allowance. In the long run z will be lower by about 45 percent.The tax adjusted Q on the other hand increases immediately by 16.5 percent and by 11 percent in the following year. In 1986 though the negative effect of lower depreciation allowances will dominate the positive effect of the tax cut and tax adjusted Q will fall by 10 percent with its value being in the steady state lower by about 14 percent.

Since investment responds with a one period lag to changes in Q, it will remain unchanged in 1984. In 1985 and 1986 it will increase by about .8 percent and it will subsequently fall in 1987 by .4 percent as Q falls. In 1988 it will fall by .67 percent and in the long run it will be lower by .7 percent. The relatively small response of investment is due to the large value of theadjustment cost parameter. To test the sensitivity of our results to the value of this parameter, the model was solved with the adjustment cost parameter taking the value of 170. Investment this time rose by about 1.5 percent in 1985, but it fell by 1.6 percent in the long run.

6. CONCLUSIONS

In this paper we have tried to investigate the effects of the 1984 Tax Reform on private nonresidential investment in the U.K. The marginal valuation of capital will increase as after tax profits increase making investment more profitable. At the same time the abolition of the first year allowance and the cut in the corporation tax rate will decrease the tax adjusted Q making investment less profitable. During the transitional period, the positive effect will dominate and investment will increase. Once however the reform is completed , investment will fall, in the long run being lower by .7 percent . A number of aspects that we ignored in this paper should be addresed in the future. That includes the treatment of tax exhaustion, the separation of gestation from expectational lags and the separate treatment of buildings and plant and machinery.

DATA APPENDIX

I is private nonresidential gross investment at 1980 prices. Source: Blue Book (BB).

IN is private nonresidential investment at current prices. Sources:BB.

p^k is the investment deflator

K is beginning of period private nonresidential gross capital stock at 1980 prices. Source:BB

OD is gross dividends on ordinary shares, of Companies and Financial Institutions.Since 1973 the series have been grossed up . Source: BB

PD is gross dividends on ordinary shares, of Companies and Financial Institutions. Since 1973 the series have been grossed up . Source: BB

PA is gross dividends on ordinary shares, of Companies and Financial Institutions. Since 1973 the series have been grossed up . Source: BB

DB is the inerest payments on long term debt. Source: BB.

r^e is the dividend yield on the FT – Actuaries Index of 500 Industrial Ordinary Shares. Source: Annual Abstract of Statistics (AAS).

r^p is the dividend yield on preference shares. Source: AAS

r^d is the FT – actuaries redemption yield on 20 year industrial debentures and loan stock. Source: AA.

NK is beginning of period net capital stock at current replacement cost of Companies and Financial Institutions. Source: BB.

BIN is the beginning of period book value of inventories of Companies and Financial Institutions. Source: BB.

z^d is the writting down allowance. Source: Dinenis(1985) and Inland Revenue Statistics, various issues.

z^g is the investment grant. Source: Dinenis(1985) and Inland Revenue Statistics, various issues.

z^a is the fiirst year allowance. Source: Dinenis(1985) and Inland Revenue Statistics, various issues.

z^i is the investment allowance. Source: Dinenis(1985) and Inland Revenue Statistics, various issues.

τ is the corporation tax rate.

r^b is the after tax nomonal interest rate on British government consols. Source: AAS·

B is the discount factor used to calculate the present value of depreciation allowances and it is defined as $B = 1/(1 + r^b)$.

REFERENCES

ABEL, A.B.: "Dynamic Effects of Permanent and Temporary Tax Policies in a q Model of Investment". *Journal of monetary Economics* 9, (1982), 353–373.

AUERBACH, A.J.: "Tax Reform and Adjustment Costs : The Impact on Investment and Market Value" *NBER Working Paper No* 2103 (1986)

DINENIS, E.: "Adjustment Costs, Q,Taxation and Investment in the U.K". LSE, CLE *Discussion Paper No* 235, (1985)

EDWARDS, J.: "The 1984 Corporate Tax Reform". *Fiscal Studies*, 5, (1984), 30–40

FISHER, P.G., HOLLY, S. and HUGHES HALLET, A.J.:"Efficient Solution Techniques for Dynamic Rational Expectations Models". *Journal of Economic Dynamics and Control.* 6 (1986).

HAYASHI, F.: "Tobin's Marginal Q and average Q: A Neoclassical Interpretation", *Econometrica*, Vol 40 (1982)

HMSO Note by HM Treasury, Appendix 10 to *Fourth Report from the Treasury and civil Service Committee*: The 1984 Budget. Session 1983–84, HC 341 (1984)

SUMNER, M.T.: "Investment and The 1984 Budjet: An Interim Assessment" *Oxford Bulletin of Economics and Statistics* 48, 4 (1986)

SEIERSTAD, A. and N. SYDSAETER : " Optimal Control Theory With Economic Applications ". *North Holland* (1987)

Taxation of Capital Income in the Federal Republic of Germany

W. Leibfritz

Ifo-Institute for Economic Research, Munich

1. Introduction

During the 1970s until the early 1980s in most industrial countries a major objective of tax policy and fiscal policy was to stimulate economic growth by special investment promotion measures. Various instruments were used as investment tax credits, favourable depreciation allowances etc. The combination of high statutory tax rates and favourable depreciation allowances provided a strong impulse to business investment. This philosophy has changed and the main objective of present tax reforms is to reduce statutory tax rates and to broaden the tax base. By that means the tax system is supposed to become fairer, simpler and less vulnerable to ad hoc changes.

This new approach of tax policy started in those countries where tax treatment of investment has been especially favourable, namely in the United Kingdom, and in the United States. The United Kingdom abolished its first year depreciation of 100 % and reduced the corporate tax rates substantially. The United States ended their system of investment tax credits, reduced their depreciation allowances and reduced also the statutory tax rates of corporate tax and personal income tax.

Opinion about this new approach with less favourable tax treatment of investment and lower tax rates is divided.

On the one hand this new system is praised as being less distortionary between different investment goods, economic sectors and sources of finance. Such distortionary effects were proven by empirical studies, based on model calculations as in section 2. It is argued that with a less distortionary tax system productivity of the capital stock will improve.

On the other hand the ending of investment promotion is critisized because it raises the overall burden on capital formation and reduces the growth of the capital stock. In West Germany during the past years the tax treatment of investment has been improved various times due to more favourable depreciation allowances, but investment promotion was - apart from cyclical measures - never as pronounced as in the United Kingdom or in the United States.

The recent tax reform plan is also oriented towards the "new approach" of lower tax rates and a broader tax base. However the reduction of the statutory corporate tax rate will be smaller as compared to the United Kingdom or the United States. But general depreciation allowances will not be cut.

The following analysis of capital taxation is divided into three parts. In the first part the integrated capital taxation model which includes taxes on the level of investors and on the level of savers is updated. Here the questions are posed:

- How did the real effective marginal tax rate on capital income develop in the past and how will it change by the forthcoming tax reform?

- Which are the major changes as compared to the previous King/ Fullerton study (4)?

In the second part a time series is calculated for user capital cost and its components in the manufacturing sector for the past 25 years. Here the questions are posed:

- How did capital cost develop?

- Which were the major explanatory factors for this development?

In the third part an attempt is made, to assess the impact of taxes and of capital cost on business investment. Here the results of various business surveys are discussed which have been carried out by the Ifo Institute during the past years.

2. The Integrated Capital Taxation Model

2.1 General Remarks

This model measures the real effective marginal tax burden on the private capital formation process. It was originally developed and applied to four countries: the United States, the United Kingdom, Sweden and West Germany in King and Fullerton (4). Most recently this model has also been applied to other countries. The model is well described in the previous study. For the German version see also Leibfritz (6). Therefore table 1 shows only the main factors which influence the final result.

- As taxes are included at the investment <u>and</u> at the saving level not only is business taxation important (here the model is confined to taxation of corporate enterprises) but also all taxes on income and wealth of private and institutional savers are important.

- As tax treatment may differ between the various investment goods (equipment investment, construction, stockbuilding), the various groups of savers and the different sources of business financing (debt financing, equity financing), the structure of capital formation is important.

- As the tax burden is defined as a real marginal tax rate on capital formation, the results also depend on the real interest rate and on the inflation rate, taking into account the sensitivity of the tax system to inflation.

Clearly, such a broad approach is quite ambitious. It is also clear, that the results are only valuable if the parameters concerning the tax system and the structure of capital formation are realistic. For some parameters the empirical finding poses no problem as the tax code and tax statistics and other statistics provide the necessary information.

But for other parameters there is a good deal of uncertainty in which great estimation problems exist. This is especially the case for variables such as economic depreciation, debt- equity ratios, ownership of capital, average marginal income tax on capital income, not to mention the problem of a correct measurement of the real interest rate.

In international comparisons these empirical problems may be even bigger, as available statistics and studies needed to be relied upon may not be comparable. When the above mentioned international comparison was made some years ago, much time and research was spent to find the correct and comparable parameters. But this does not mean that improvements are not still possible. Furthermore, as tax systems are bound to change quite often, such calculations have to be updated from time to time. Yet if this updating is done separately for the individual countries, we cannot be sure that the international comparibility still holds. In the following the results of an updated model calculation for Germany are presented.

2.2 Empirical Results

2.2.1 Overall Results

The tables 2 to 4 show the tax parameters and the weights and table 5 and diagram 1 the effective marginal tax rates under this approach for the years 1965, 1970, 1980, 1987 and 1990 i.e. after the tax reform measures. The effective tax rates are shown for the fixed P-case (constant real gross return of 10 %) and the fixed R-case (constant real interest rate of 5 %). With the latter case the level of the tax rate is higher.

In 1965 and 1970 the overall effective tax rates have been at levels of about 50 % (fixed P-case) and 70 % (fixed R-case). They are declining since them.

In the year 1987 these rates are at levels of about 39 % (fixed P-case) and 64 % (fixed R-case). The main reasons for this decline were improvements of depreciation allowances for both machinery and buildings. The tax reform of 1990 will bring about a further moderate decline of the effective tax rates. Concerning capital taxation especially three elements of this reform have to be taken into account: the reduction of the marginal tax rates for personal income, the reduction of the corporate tax rate for retained earnings and the introduction of a 10 % withholding tax on certain interest income. A summary of the various measures is given in table 6 and the new personal income tax schedule is shown in diagram 2. While the reform of the personal income tax will reduce the overall effective tax rate by about 2 percentage points (fixed P-case), the reduction of the corporate tax rate will increase it by about 1 percentage point. But this is not the total effect of the withholding tax. In the model calculation only the withholding tax paid by insurance companies has been considered because here the coverage of te tax law will be extended. The withholding tax which has to be paid by banks (about 75 % of total withholding tax revenues) will reduce tax evasion. But this has not been taken into account in the model explicitly, because in the base case calculation, the evasion has been neglected. The sensitivity of this assumption is shown in Leibfritz (6).

2.2.2 Structure of Tax Rates

Due to this model calculation the effective tax rates are lower for investment in machinery and equipment and in construction as compared to inventories. This gap has been widened in recent years due to the improvement of tax depreciation while stock accounting has been left unchanged. Concerning sectors, the effective tax rate is lower in the commercial sector as compared to manufacturing and other industry. The main reason is the difference in the debt/equity ratios. This ratio is relatively high in the commercial sector.

The fact that the present tax system is highly distortionary is most striking when effective tax rates on the various sources of finance are regarded. In all years debt financing was much less burdened than financing by new share issues and retained earnings. While for all three sources of finance the effective tax rate declined in 1987 as compared to 1980, for debt financing there is now a negative sign which implies subsidizing. The main reasons for these structural differences in effective tax rates have been explained in detail in the previous studies. It has been shown that the preferential treatment of debt finance is due to the possibility of the firm to deduct nominal interest payments when at the same time depreciation allowances are favourable and - as mentioned above - these allowances have been improved in recent years. The advantage at the level of the firm is not compensated at the level of the savers as their marginal income tax rates are on average lower than the corporate tax rate which is applied for the deduction of interest payments.

In addition in contrast to equity financing debt financed investment is not liable to the corporate net wealth tax.

Between 1970 and 1980 with the reform of the corporate tax, the (full) imputation system was introduced. This imputation reduced the effective tax rate for new share issues as compared to retained earnings because the average marginal income tax rate of capital owners is lower than the corporate tax rate for retentions. This implies that the opportunity cost of keeping the money in the firm (retentions) is higher than the opportunity cost of raising new capital (new share issues). With the reduction of the corporate tax rate for retentions in 1990 this difference will be reduced only marginally as the marginal tax rate for personal income will also be reduced.

2.2.3 Sensitivity of Inflation

Diagram 3 shows the overall effective marginal tax rates in the fixed P-case for inflation rates varying from zero to 10 %. As shown already in the earlier publication (4) overall marginal tax rates in Germany are rather insensitive to the rate of inflation. This is certainly the case in the range of inflation which has been relevant for Germany during the past.

2.2.4 International Comparison

When the model calculation was first used for an international comparison, the overall effective marginal tax rate was higher in West Germany as compared to all other countries, namely United Kingdom, the United States and Sweden. For the fixed P-case the tax rate was about 48 % in Germany, about 37 % in the United States, about 36 % in Sweden and about 4 % in the United Kingdom. In these calculations some general investment grants (regional investment promotion) were included.

In the meantime some important tax measures have been taken in the various countries. As mentioned above the United Kingdom and the United States have endend or reduced their investment promotion schemes and have reduced the corporate tax rate substantially. In the United States the marginal income tax rate has also been reduced. The effects of these tax measures are discussed in the other papers.

Taken as a whole it can be expected that after the tax reforms which have been introduced or which are prepared at present in the various countries the effective marginal tax rates for the countries which have been analysed in 1984 (United States, United Kingdom, Sweden, West Germany) will be in the range of about 30-40 % (fixed P-case), a range which is much smaller than it was before. This shows that most recent and forthcoming tax reforms tend to equalize the previous differences in effective tax rates on capital formation.

But given the high sensitivity to the various assumptions and given the complexity of the model calculations as suggested above these international comparisons should be coordinated again. Otherwise comparibility might not be guaranteed.

3. The Development of the Cost of Capital in the Manufacturing Sector during the past 25 Years

Here we apply a user cost of capital approach as originally defined by Hall/Jorgenson with some refinements. The most important elements of the capital cost are the real interest rate, economic depreciation, the tax advantage of depreciation allowances and other investment promotion as investment grants. Diagram 4 shows the development of capital cost for equipment investment for the past 25 years. Furthermore a tentative projection is shown until 1990 when the new tax reform will be implemented. This projection is based on a "no change" assumption for non tax factors, in order to outline the effect of the tax reform. The diagrams 5 and 6 show the development of the major elements of capital cost.

For equipment investment (diagram 4) it is shown that

- total capital cost including investment grants fluctuated around a rather flat trend with four peaks in the years 1966, 1973/74, 1981 and 1984 and seven troughs in the years 1963, 1969, 1971/72, 1975, 1978, 1982/83 and 1986/87. With the 1990 tax reform capital cost will increase (due to a lower corporate tax rate which leads to a decline of the tax advantage of depreciation),

- all four peaks were strongly influenced by the development of interest rates (diagram 5), but cyclical restrictive tax measures also contributed to one peak (1973),

- most of the troughs were due to the development of interest rates, but two of the troughs (1975, 1982/83) were also influenced by expansionary policy measures (temporary investment grants),

- the extension of investment grants for special purposes (as regional aids, research and development, sectoral aids etc.) helped to avoid a rising trend of capital cost. This is shown by the development of capital cost excluding investment grants (diagram 4),

- during most of the 1960s and 1970s the tax system was somewhat less favourable to investment than at the beginning of the 1960s but since the 1980s it is more favourable. The main reason was the improvement of depreciation allowances (see diagram 6).

For business construction it can be shown that

- cyclical fluctuations of capital cost for construction were similar as for equipment investment which is not surprising as both were influenced by the development of interest rates and by cyclical policy measures,

- the long-term trend of capital cost including investment grants was declining and if investment grants are excluded it was rather flat.

A major reason for this difference to capital cost of equipment investment was the different development of economic lifetimes which were assumed to remain unchanged, whereas economic lifetimes for machinery and equipment were reduced. A further reason was the substantial improvement of depreciation allowances (especially since the mid 1980s).

4. Capital Cost and Investment Activity

4.1 General Remarks

Diagram 7 compares the development of investment in the manufacturing sector with the development of capital cost and other supply factors as profits and also with demand as indicated by capacity utilization. Although an adverse impact of capital cost on investment is noticeable, this supply factor looses its explanatory power in periods with cyclically weak demand. This result is also confirmed by econometric studies which show that business investment in Germany is influenced by supply <u>and</u> demand factors.

In these studies various supply parameters such as nominal or real interest rates, average rates of return and tax burden indicators were used alternatively in isolation and together with a demand parameter, with and without time-lags (Leibfritz/Meurer (5), Gerstenberger (2)).

In the macroeconomic model of the Ifo Institute, which is used for short term forecasting, investment is also explained by demand and supply variables.

Due to this model, investment and GNP can be increased by income tax reductions and also by tax measures which reduce capital cost. While income tax reductions have strong effects on domestic demand which then leads to higher business investment, capital cost reductions increase investment directly, but here, due to the model, the impact on economic growth is smaller as compared to income tax reductions. This may to some extent be due to the nature of the model which considers demand effects more than supply effects.

4.2 Business Survey Approach

4.2.1 Investment Barriers in the German Manufacturing Sector

In the context of the regular investment survey of the Ifo Institute, a special study was made about the main investment barriers in the manufacturing sector (see Uhlmann (8)). The study covered the years 1973 to 1977. The questionnaire was answered by 1,935 firms, which account for about one third of total investment in the manufacturing sector. Various types of investment (capital widening, capital deepening, environment protection) and a large number of possible dampening factors (21) were considered and firms had the possibility to combine them. A possible negative impact of taxes on investment was considered by various explanations which could be marked: "insufficient depreciation allowances", "high tax burden", "uncertainty about future tax burden". About 20-30 % of the firms expressed a negative impact of these factors, but only for a few of them were they the most important ones. The most important restraining factors for investment were demand factors, such as "underutilization of capacity", "low future economic growth" and "uncertainties about future economic growth".

4.2.2 Effects of the Measures for Cyclical Stabilization Purposes

Apart from this general study about the determinants of business investment, business surveys were also made to evaluate the impact of the various temporary tax measures for stabilization purposes (see table 7) (see Gerstenberger (3), Apeloig (1)).

In 1970 restrictive measures were not strong enough to exert the desired dampening effect on investment. But in 1973 when restrictive measures were much stronger and when tax policy was supported by monetary policy, the investment boom was dampened substantially, so that shortly thereafter, expansionary measures were necessary. This was the period of stop and go policy. The expansionary measures in 1974/75 had a favourable impact on investment, as they prevented a further downswing. But to a good extent investors only changed the timing and after the temporary measures were abolished, investment weakened. Nevertheless the resulting smoothing of the cyclical fluctuation can be viewed as a success. The expansionary measures of 1982/83 can be viewed similarly, but their size and their effects were smaller than in 1974/75.

4.2.3 Tax Effects on Investment due to an International Business Survey

In the context of its regular international business survey on the economic development (ESI) in 1983, the Ifo Institute posed special questions about the major investment problems and about the efficiency of tax measures (Röthlingshöfer (7)). These questions were answered by 482 businessmen in 50 countries.

The biggest problems for business investment in these countries were seen at that time as insufficient demand, insufficient rate of return, and high interest rates. High tax rates were noticed by about one fifth of the participants and insufficient depreciation allowances by about 5 %. In most countries special investment promotion schemes were available, but opinion about its efficiency was divided, with a (small) majority rating them as inefficient. There were some differences between countries and country groups but by and large the results of our other studies were confirmed that both demand and supply factors play a role in investment activity, with demand factors seen as more important. Among the supply factors, the influence of the tax system and of investment promotion schemes is noticable, though these factors do not have the most important influence.

4.3 Some Remarks on Possible Effects on the Forthcoming Tax Reform on Investment

As shown above with the tax reform package in 1990 capital cost may not change substantially. Measured with the traditional user cost of capital approach capital cost will increase slightly while the integrated tax model as described in part 1 shows a moderate decline in the marginal tax rate. The most important impact on the economy will come from the reduction of the personal income tax. As experience has shown, the income effects on consumer demand and the demand effect on business investment will be important even if, given the restrictive public spending policy, only the net impact on demand is considered.

In 1985/86 business investment recovered due to an increase in exports, domestic consumption, lower capital cost and higher profits. During 1987 growth of business investment remained moderate and it is expected to decelerate further during 1988. This will be the case although consumer demand will be bolstered by the income tax cut. There are still high uncertainties about the development of exports and with an export share of 50-60 percent in key branches of industry business investment has always been strongly influenced by export expectations. In 1990 when the main part of the tax reform package becomes effective, business investment should be stimulated more. The main impact of this tax reform will again not be a general improvement of business supply conditions but rather a stabilization of overall domestic demand, due to the reduction of the personal income tax. It is possible that business climate also improves somewhat due to psychological factors, stemming from the reduction of statutory income and corporate tax rates. But on the other hand business taxes in Germany are still felt to be rather high by international standards. The pressure will therefore remain for the government to go on with reforming the tax system by a more pronounced "supply side step" of a further reduction in business taxes.

REFERENCES

(1) APELOIG, SH., GERSTENBERGER, W., GLUCH, E., LAUMER, H., NEUMANN, F., STÄDTLER, A., AND STANGLMAYER, M. (1984): Effects of the Investment Grant (in German), Ifo-Studien zur Finanzpolitik Nr. 32.

(2) GERSTENBERGER, W. (1980): Demand and Factor Prices as Determinants of Investment (in German), in: Ifo-Studien 1980/1-2.

(3) ————(1977): Lessons about Stabilization Policy Measures During the Past Decade (in German), in: Beihefte der Konjunkturpolitik, Heft 23, 1977.

(4) KING, M.A., AND FULLERTON, D. (1984): The Taxation of Income from Capital: A Corporative Study of the United States, the United Kingdom, Sweden, and West Germany, Chicago and London: The University of Chicago Press.

(5) LEIBFRITZ, W./MEURER, C. (1984): Investment Tax Measures in International Comparison (in German), Schriftenreihe des Ifo-Instituts für Wirtschaftsforschung Nr. 117, Verlag Duncker & Humblot Berlin-München.

(6) LEIBFRITZ, W. (1986): Capital Taxation and Promotion Measures in the Federal Republic of Germany (in German), Ifo-Studien zur Finanzpolitik Nr. 36.

(7) RÖTHLINGSHÖFER, K.CH. (1983): No Optimism in Industrial Countries, Ifo-Schnelldienst Nr. 13.

(8) UHLMANN, L. (1981): Investment Behaviour of Industry due to Business Surveys (in German), Schriftenreihe des Ifo-Instituts für Wirtschaftsforschung Nr. 105, Bd. 2.

Table 1 Influences on the Effective Tax Rate due to the Capital Taxation Model

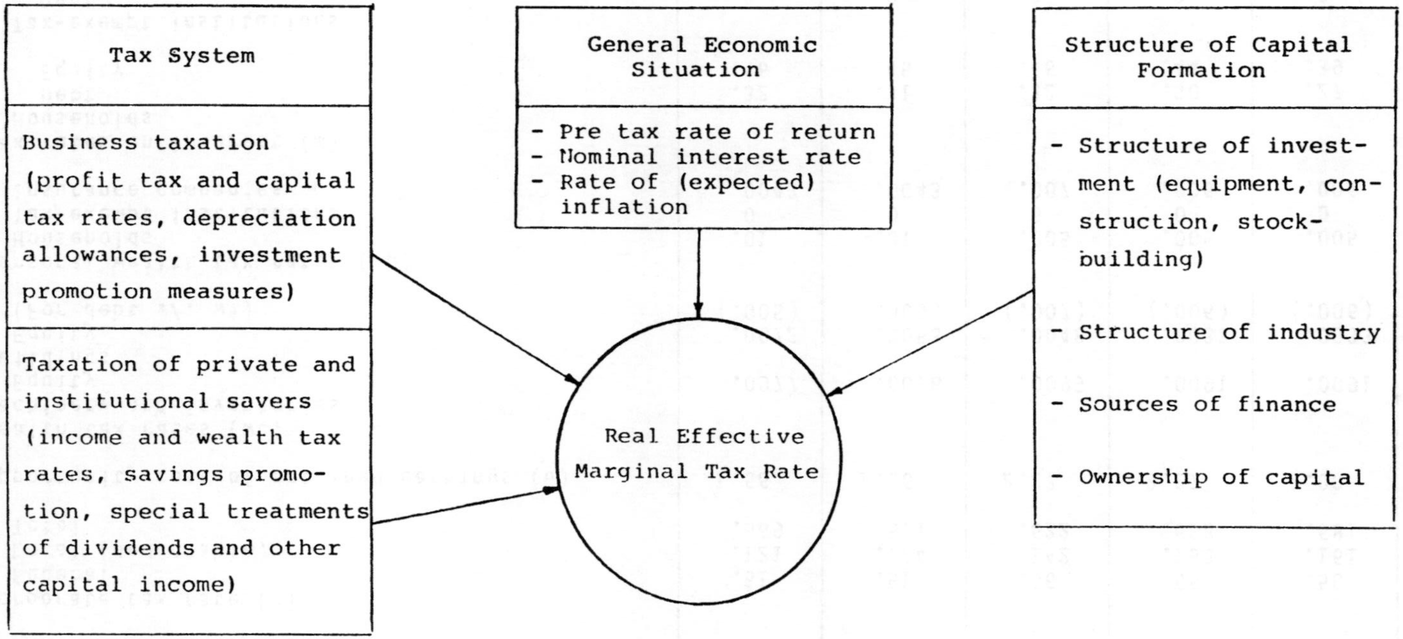

Table 2

Tax Parameters (1)

	1965	1970	1980	1987	1990
Corporate tax rate (τ)					
Federal	.51	.51	.56	.56	.50
Local (deductable)	.121	.124	.142	.153	.161
Total	.569	.571	.622	.628	.581
Opportunity cost of retained earnings (θ)	1.56	1.56	2.27	2.27	2.00
Wealth tax rates (wc)					
Machinery and inventories					
Equity	.0077	.0078	.0099	.0091	.0091
Buildings					
Equity	.0072	.0053	.0048	.0033	.0033
(For debt ./. w*)	(.005)	(.005)	(.007)	(.006)	(.006)
Personal wealth tax rates (wp)					
Households	.01	.01	.005	.005	.005
Tax-exempt institutions	0	0	0	0	0
Insurance companies	.0043	.0043	.007	.006	.006
Tax rates on interest (m)					
Households					
Debt	.32	.31	.32	.30	.27
Equity	.46	.45	.46	.44	.39
Tax-exempt institutions					
Debt	0	0	0	0	0
Equity	.246	.246	.40	.40	.40
Insurance companies					
Debt and equity	.09	.09	.09	.09	.14

Table 3

Tax Parameters (2)

	1965	1970	1980	1987	1990
Depreciation rates (δ)					
Machinery					
Manufacturing	.145	.152	.174	.179	.182
Other industry	.213	.217	.230	.235	.238
Commerce	.217	.225	.238	.250	.260
Buildings					
Manufacturing	.037	.037	.037	.037	.037
Other industry	.030	.030	.030	.030	.030
Commerce	.022	.022	.022	.022	.022
Economic lifetimes (L)					
Machinery					
Manufacturing	13.8	13.2	11.5	11.2	11.0
Other industry	9.4	9.2	8.7	8.5	8.4
Commerce	9.2	8.9	8.4	8.0	7.7
Buildings					
Manufacturing	53.4	53.4	53.4	53.4	53.4
Other industry	67	67	67	67	67
Commerce	92	92	92	92	92
Rates of declining balance (machinery) (B)[a)]	2.0	2.0	2.5	3.0	3.0
Tax lifes (Lt)					
Machinery					
Manufacturing	7.4	7.0	6.5	6.2	6.0
Other industry	5.9	5.7	5.0	4.9	4.7
Commerce	4.9	4.7	4.3	4.0	3.9
Buildings					
Manufacturing	46.4	46.4	46.4	25	25
Other industry	48	48	48	25	25
Commerce	50	50	50	25	25
Rates of tax depreciation					
Machinery					
Manufacturing	.20	.20	.25	.30	.30
Other industry	.20	.20	.25	.30	.30
Commerce	.20	.20	.25	.30	.30
Buildings					
Manufacturing	.022	.022	.022	.04	.04
Other industry	.021	.021	.021	.04	.04
Commerce	.02	.02	.02	.04	.04

a) With maximum rates of 20 %, 25 % and 30 % respectively.

Table 4

Weights

	Manufac-turing	Other industry	Commerce
Proportion of capital stock			
Machinery	.2894	.0244	.0359
Buildings	.2594	.0256	.0976
Inventories	.2130	.0048	.0499
Proportion by source of finance			
Debt	.504	.42	.546
New share issues	.076	.076	.076
Retained earnings	.420	.504	.378
Ownership shares			
Debt			
Households	.73	.73	.73
Tax-exempt institutions	.20	.20	.20
Insurance companies	.07	.07	.07
Equity			
Households	.758	.617	.751
Tax-exempt institutions	.186	.298	.085
Insurance companies	.056	.085	.060

Table 5

Development of the Effective Marginal Tax Rate [a)]

	Fixed P-case					Fixed R-case				
	1965	1970	1980	1987	1990	1965	1970	1980	1987	1990
ASSET										
Machinery	45.9	47.0	46.1	39.1	37.5	72.1	72.2	70.6	66.6	62.1
Buildings	54.7	51.6	44.6	29.8	29.2	74.1	71.5	64.8	50.8	46.1
Inventories	58.0	57.6	55.1	53.7	51.3	75.5	74.9	72.1	70.7	66.7
INDUSTRY										
Manufacturing	53.0	52.4	49.6	41.2	39.5	74.4	73.5	70.3	64.9	60.4
Other industry	57.2	56.2	50.9	42.8	41.5	75.7	74.7	71.0	65.7	61.7
Commerce	48.9	47.1	40.2	31.1	30.7	70.5	68.6	62.3	55.2	50.9
SOURCE OF FINANCE										
Debt	20.6	19.0	12.0	± 0.0	2.0	34.3	30.2	11.9	- 8.0	- 3.1
New share issues	83.4	82.7	64.6	56.3	55.1	90.4	89.6	80.3	74.7	70.8
Retained earnings	85.6	85.6	88.6	84.3	78.7	90.9	90.3	88.7	86.4	82.3
OWNER										
Households	63.0	61.8	57.2	48.7	45.3	85.8	84.4	78.3	73.0	66.7
Tax exempt institutions	24.0	23.9	22.0	13.9	16.5	41.6	41.3	43.6	37.5	36.5
Insurance companies	19.4	19.4	20.9	11.2	20.2	32.5	32.1	38.9	30.2	37.2
OVERALL	52.5	51.6	47.9	39.4	38.0	73.9	72.7	69.1	63.5	59.0

a) Excluding investment grants for special purposes

Table 6

The 1990 tax reform

Stage I	Loss of tax revenues in DM billion (per year)
1. Reduction of personal income tax (from 22/56 % to 19/53 % rate structure and other measures)	41.6
2. Reduction of corporate tax rate (for retained earnings from 56 % to 50 %)	2.3
3. Increase of special depreciation allowances for small sized enterprises (from 10 % to 20 %)	.5
Total reduction (1+2+3)	44.4 (of which 5.2 advanced to 1988)
Stage II (compensatory measures)	**Increase of tax revenues in DM billion (per year)**
1. Broadening of income and corporate tax base and reduction of other tax promotion measures	
- for the business sector and for professionals	3.0
- for employees	4.0
- for income tax payers in general	4.1
2. Introduction of a withholding tax on interest income	4.3
3. Reduction of special investment grants (regional aid) and other promotion measures	2.8
Total compensatory measures (Original plan for compensatory measures	18.2 19.4)
Total tax reduction net (Stage I ./. Stage II) (Orginal plan = 5 % of tax revenues or 1 % of GNP)	26.2 25.0)

Table 7 Effects of Tax Measures for Cyclical Stabilization Purposes

- Results of various Business Surveys of the Ifo Institute -

Measures	Reaction of Business due to Survey	Explanation
Restrictive measures 1970		
(temporary abolition of declining balance depreciation, 10 % repayable surcharge on income and corporate tax)	low (about 8 % of firms plan to reduce investment)	general optimistic outlook, price increases
Restrictive measures 1973		
(temporary abolition of declining balance depreciation, 10 % surcharge on income and corporate tax, 11 % investment tax)	substantial (about 60 % of firms plan to reduce investment)	strong measures, restrictive monetary policy, revaluation of currency
Expansionary measures 1974/75		
(temporary investment grant of 7.5 %)	temporary substantial (about 50 % of firms plan to increase investment but to a large part only change in timing as underlying trend of investment determinants prevails)	strong measures
Expansionary measures 1982		
(temporary grant of 10 % on additional investment)	temporary noticeable but less strong than in 1974/75 (about 25 % plan to increase investment)	measures less strong

Diagram 1

DEVELOPMENT OF THE EFFECTIVE MARGINAL TAX RATE a)

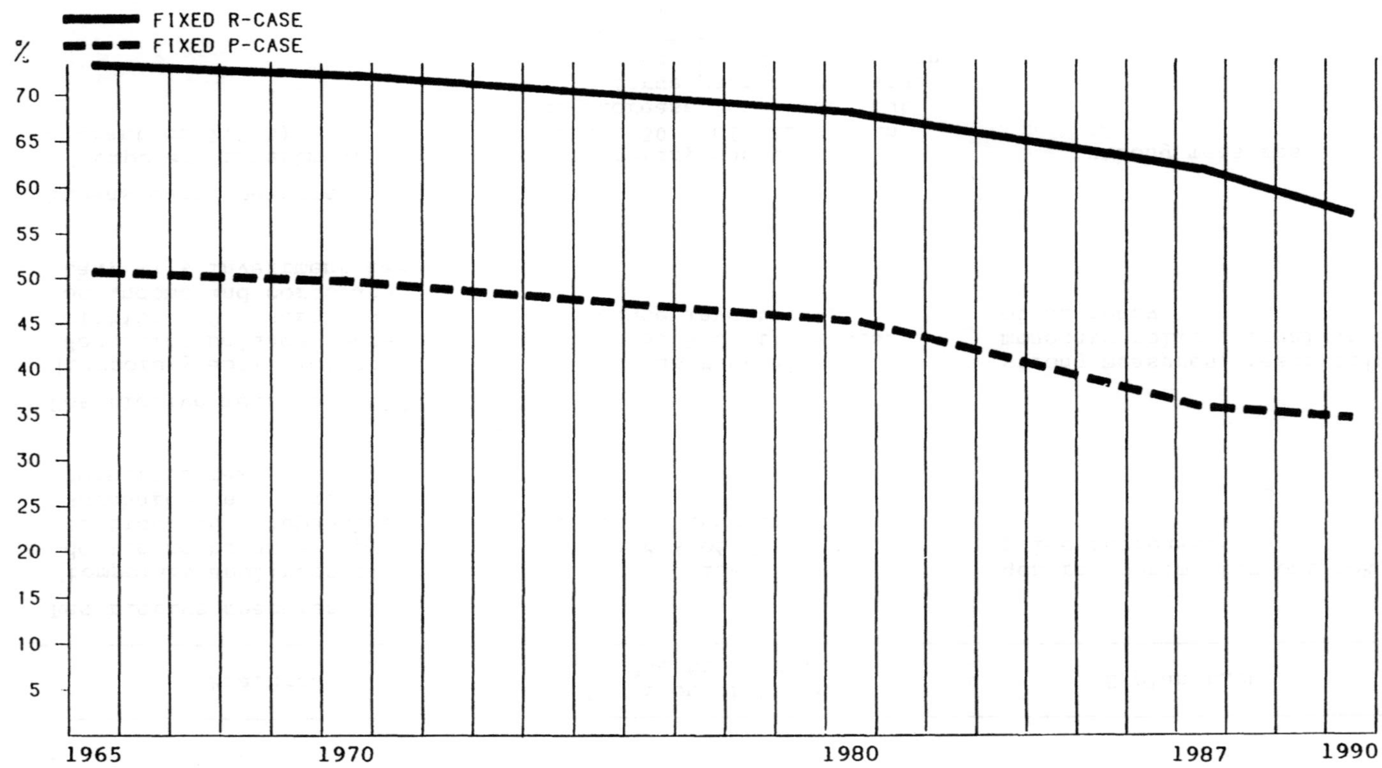

IFO-INSTITUT

a) EXCLUDING INVESTMENT GRANTS

SOURCE : OWN CALCULATIONS

Diagram 2

INTERNATIONAL COMPARISON OF INCOME TAX SCHEDULES

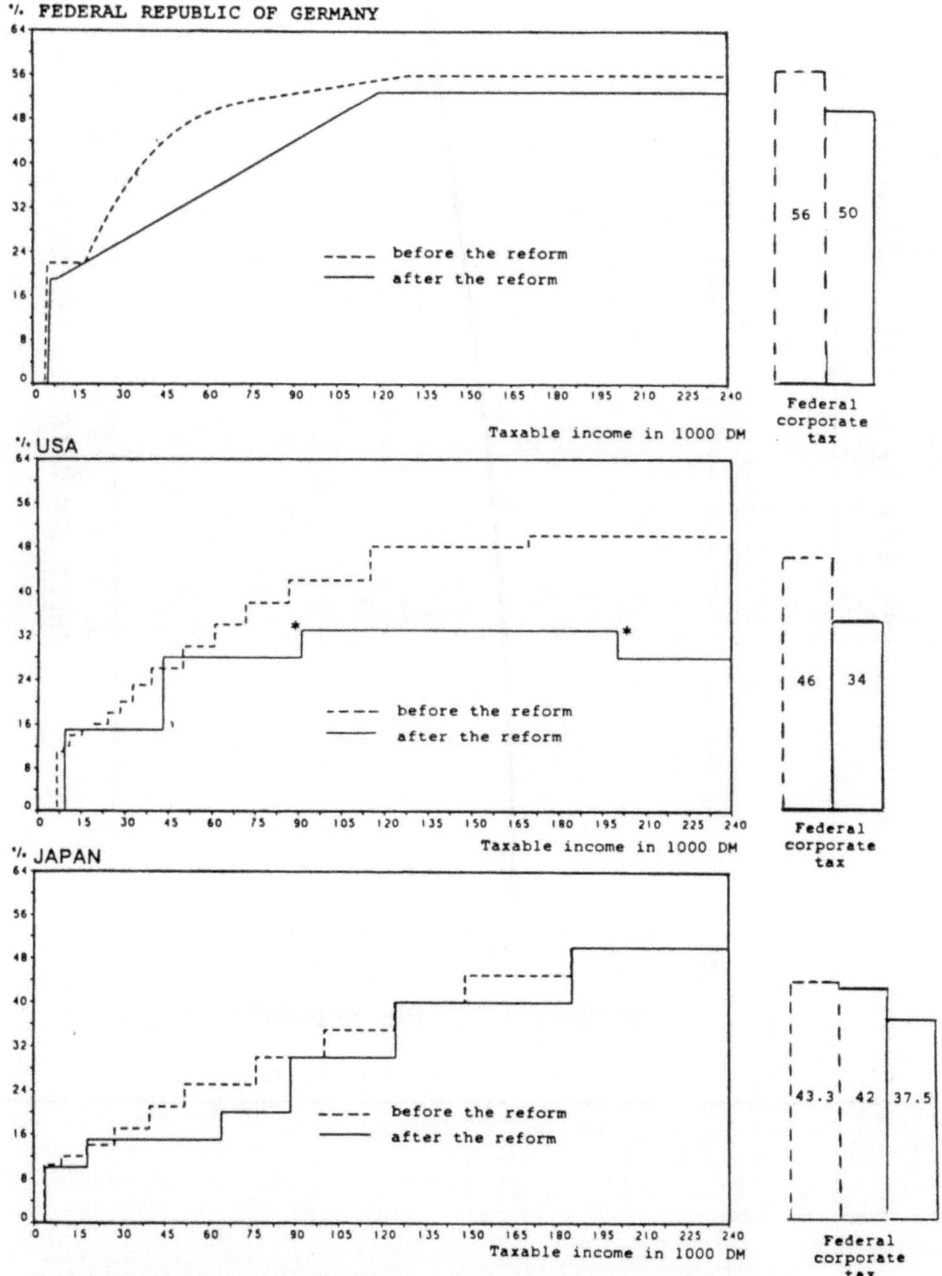

* higher marginal tax rate in order to increase average tax rate of higher income brackets to 28 %

230/87

Source: National tax laws and calculations of the IFO-Institute

Diagram 3

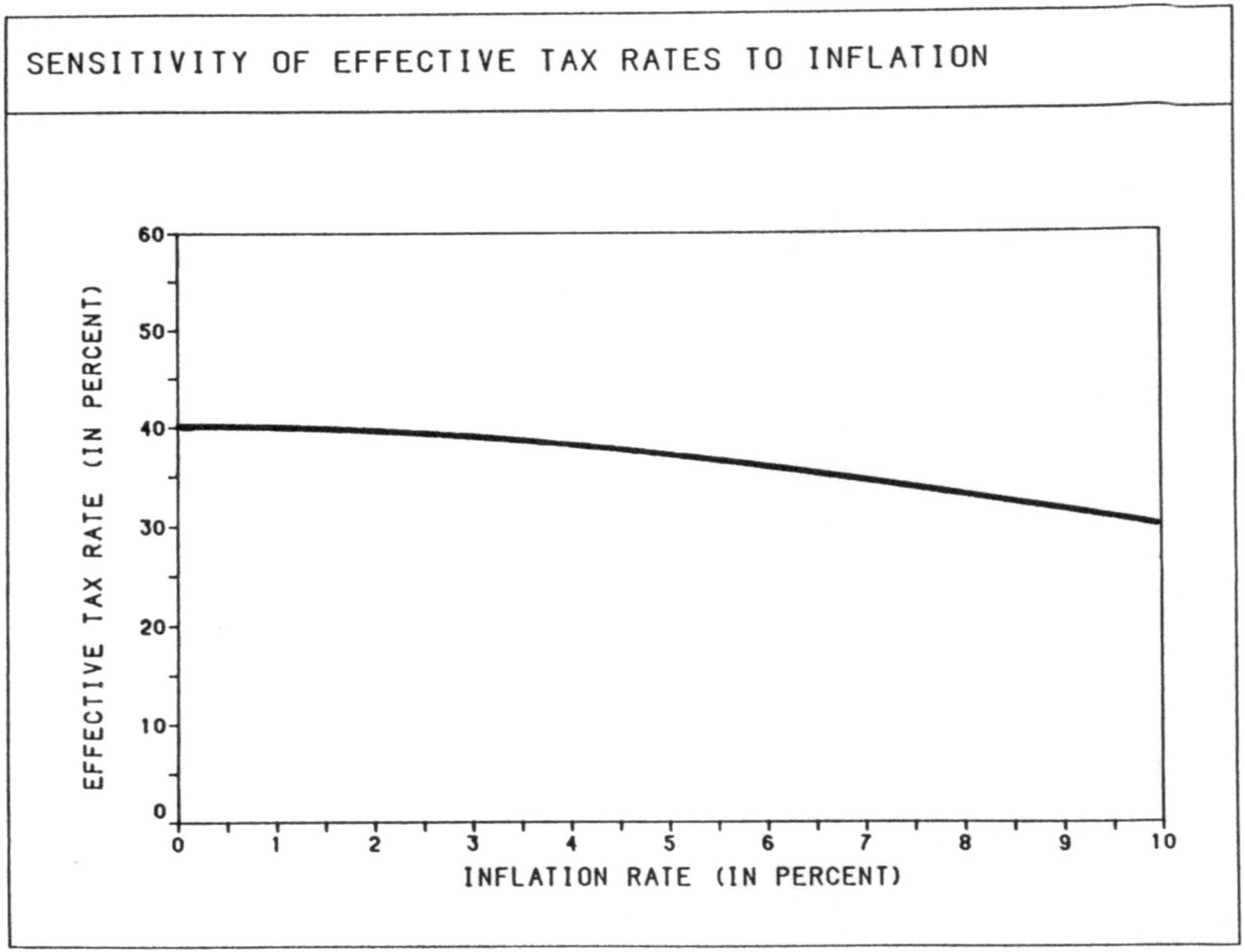
SENSITIVITY OF EFFECTIVE TAX RATES TO INFLATION
EFFECTIVE TAX RATE (IN PERCENT)
60
50
40
30
20
10
0
0
1
2
3
4
5
6
7
8
9
10
INFLATION RATE (IN PERCENT)

Diagram 4

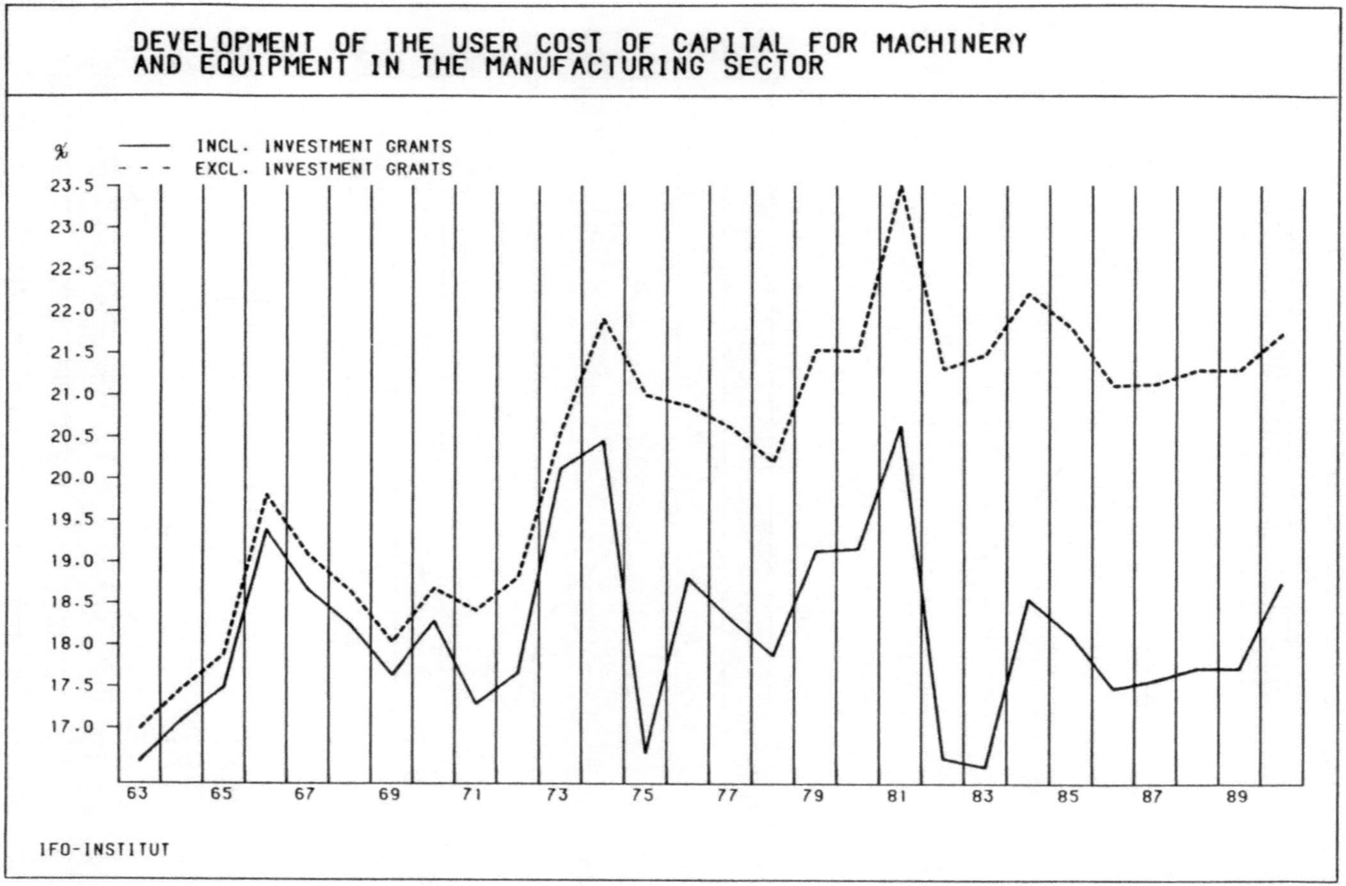
DEVELOPMENT OF THE USER COST OF CAPITAL FOR MACHINERY AND EQUIPMENT IN THE MANUFACTURING SECTOR
%
INCL. INVESTMENT GRANTS
EXCL. INVESTMENT GRANTS
23.5
23.0
22.5
22.0
21.5
21.0
20.5
20.0
19.5
19.0
18.5
18.0
17.5
17.0
63
65
67
69
71
73
75
77
79
81
83
85
87
89
IFO-INSTITUT

Diagram 5

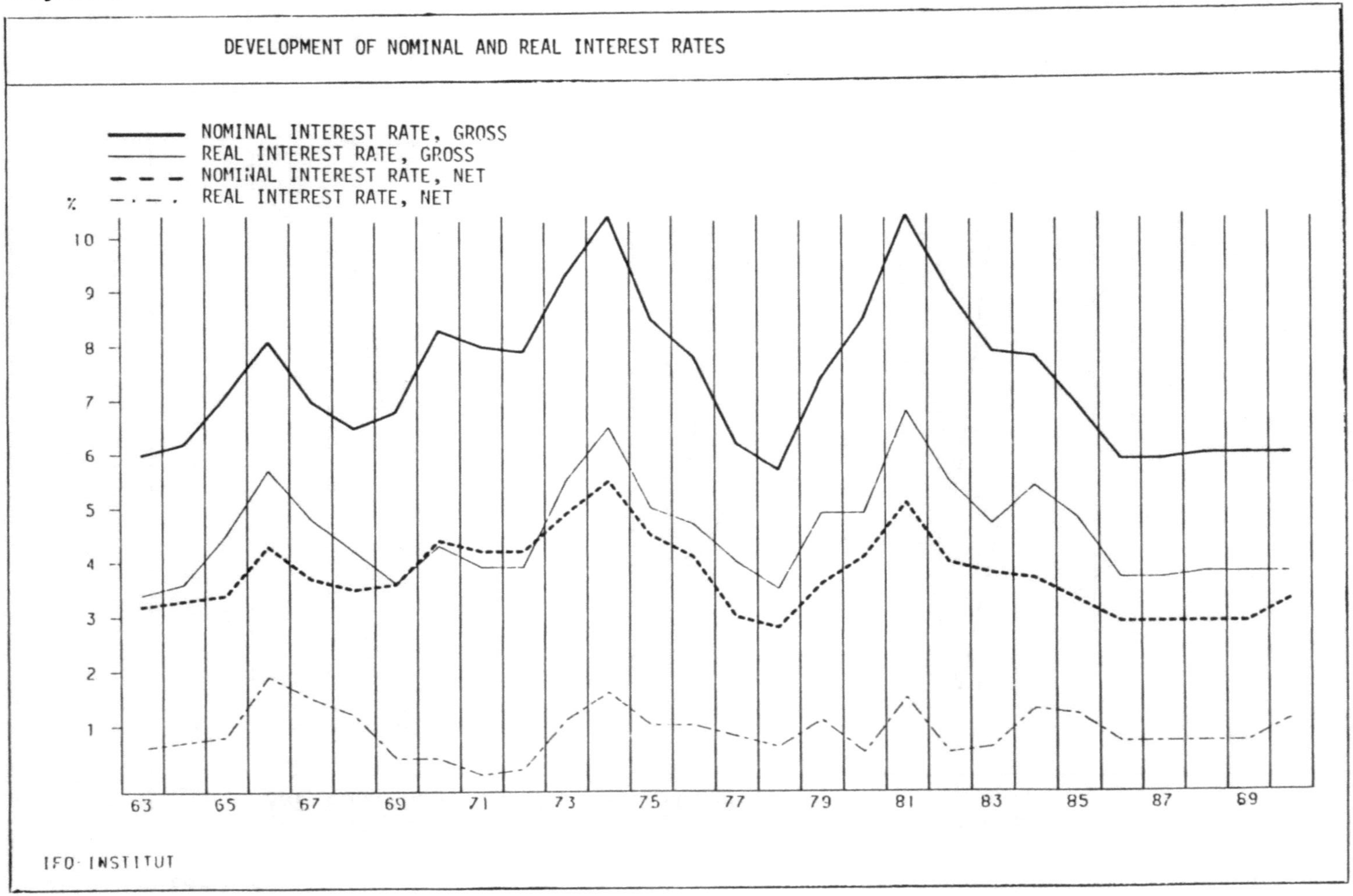
DEVELOPMENT OF NOMINAL AND REAL INTEREST RATES
NOMINAL INTEREST RATE, GROSS
REAL INTEREST RATE, GROSS
NOMINAL INTEREST RATE, NET
REAL INTEREST RATE, NET
%
10
9
8
7
6
5
4
3
2
1
63
65
67
69
71
73
75
77
79
81
83
85
87
89
IFO-INSTITUT

Diagram 6

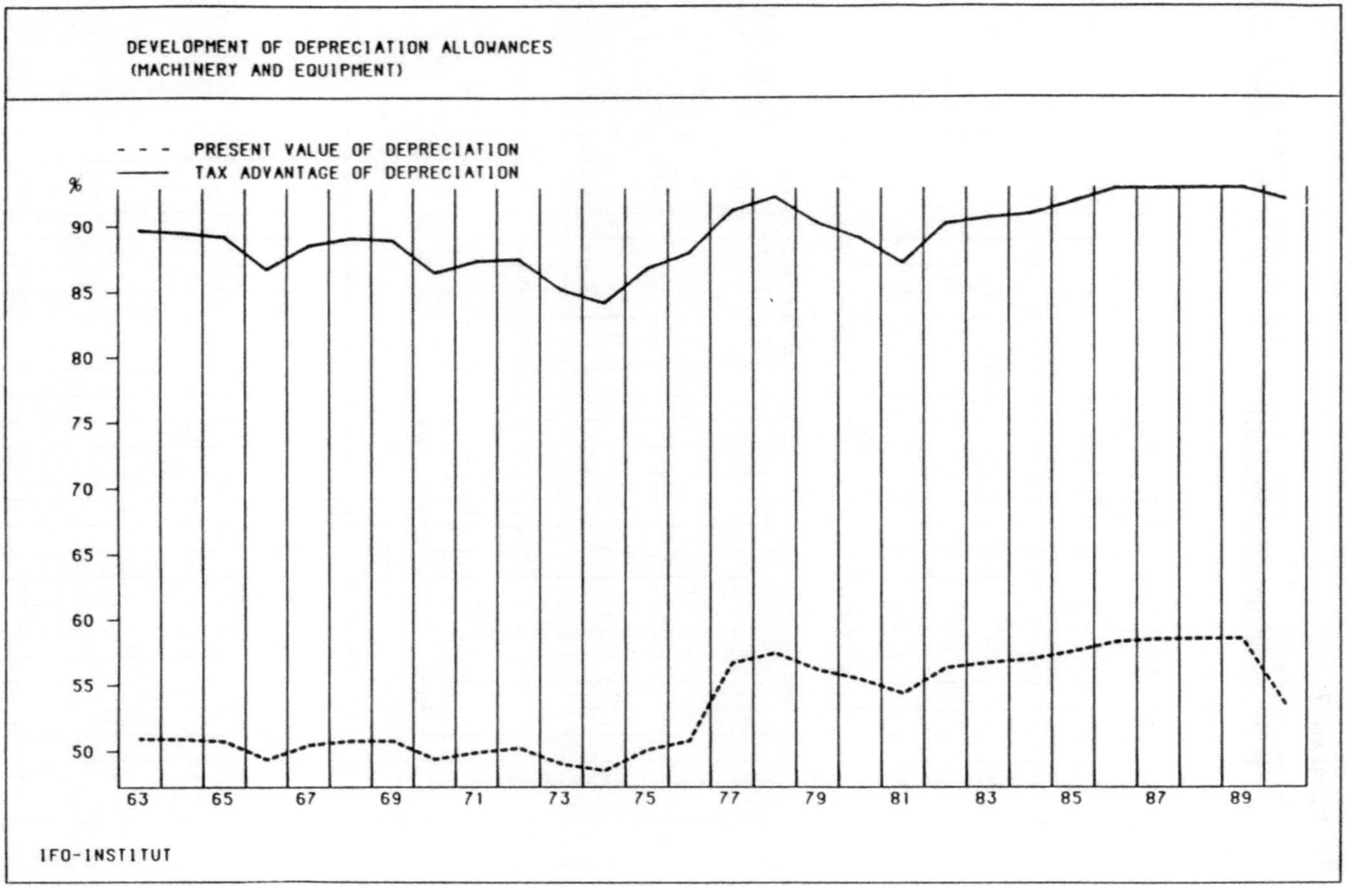
DEVELOPMENT OF DEPRECIATION ALLOWANCES
(MACHINERY AND EQUIPMENT)
- - - PRESENT VALUE OF DEPRECIATION
—— TAX ADVANTAGE OF DEPRECIATION
%
90
85
80
75
70
65
60
55
50
63
65
67
69
71
73
75
77
79
81
83
85
87
89
IFO-INSTITUT

Diagram 7

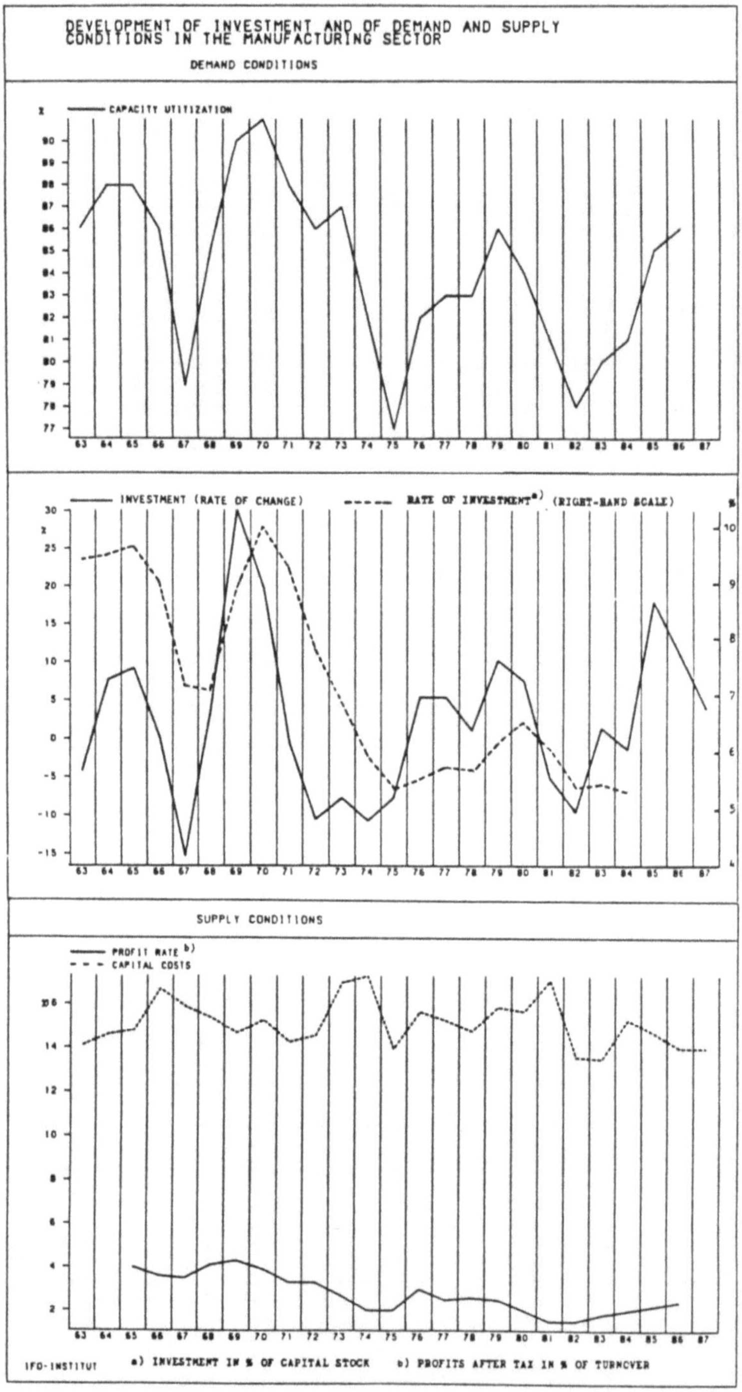

DEVELOPMENT OF INVESTMENT AND OF DEMAND AND SUPPLY CONDITIONS IN THE MANUFACTURING SECTOR
DEMAND CONDITIONS
CAPACITY UTILIZATION
INVESTMENT (RATE OF CHANGE)
RATE OF INVESTMENT a) (RIGHT-HAND SCALE)
SUPPLY CONDITIONS
PROFIT RATE b)
CAPITAL COSTS
IFO-INSTITUT
a) INVESTMENT IN % OF CAPITAL STOCK
b) PROFITS AFTER TAX IN % OF TURNOVER

Capital Mobility, the Cost of Capital Under Certainty and Effective Tax Rates in Europe

Julian S. Alworth and Wilhelm Fritz*

Bank for International Settlements, CH-4002 Basel, Switzerland

1. Introduction

The degree of integration between financial markets has increased substantially in recent years. Casual observation of simple correlations of changes in nominal interest rates, such as those shown in Tables 1 and 2 suggest that since the mid-1970s co-movements of long-term and to a lesser extent short-term nominal interest rates have tended to increase markedly, particularly during the 1980s and most noticeably amongst countries in the European Monetary System. Moreover, closer study of this question by Obstfeld (1986a, 1986b) have confirmed the basic trends implied by Tables 1 and 2 and cast doubt on an earlier empirical study by Feldstein and Horioka (1980) which suggested indirectly through simple correlations of national savings and investment that capital markets were segmented.[1]

Such a growing degree of integration means that it is no longer possible to assume that fiscal policies affecting interest rates in a single country will be without repercussions abroad and at the same time that they would have the same impact as in a closed economy. Whilst the effects of financial integration on the conduct of macroeconomic policy are well known, they have hardly been examined with respect to resource allocation and cross-country investment decisions.

The purpose of this paper is twofold. Firstly, it sets out the issues involved in defining a financial equilibrium in an open economy with a high degree of inter-country capital mobility and a distortionary tax system. We shall demonstrate that even under the simplifying assumption of certainty many alternative equilibria are possible depending on the interaction amongst interest parity conditions, the type of tax systems considered and the limitations set on tax arbitrages.

The specification of the equilibrium in the international financial markets is relevant to the second objective of this paper, namely to determine what is the appropriate measurement of effective tax rates in an open economy. As has been shown in studies of marginal effective tax rates in a closed economy, the actual values of effective rates hinge to a considerable extent on the nature of market arbitrages and of capital market equilibria which are assumed (Bradford and Fullerton (1981) and King and Fullerton (1984). In the second section of this paper we shall examine how the effective tax rates obtained using the King and Fullerton framework need to be modified in an open economy when domestic savings and investments need no longer be equal. The countries which are used as benchmarks for this exercise are five of the original EC countries plus the United Kingdom. The simulation results obtained are preliminary because of data imperfections and the difficulties in modelling many complex interactions across countries. Although they cannot be taken as firm indications of the actual impact produced by the tax changes, [2] our findings suggest that the existence of capital mobility alters the value of the effective tax rates in a significant manner. The final section draws some tentative conclusions and suggests some possible areas of future research.

In order to address these issues we have restricted the scope of our study in several respects. Firstly, only foreign portfolio investments are considered. Direct investment through multinationals is not examined because a number of questions concerning the specific tax treatment of multinationals need to be taken into account (Alworth, 1987b) and because the bulk of cross-border flows are of a portfolio nature (in particular if banking flows are considered).[3] Secondly, although the paper is concerned with the interaction of policies amongst countries, strategic issues relating to how the tax system should be designed in order to respond to foreign taxes have not been addressed. In particular, questions such as "tax wars" and the relationship between taxes and tariffs are not examined.[4] Finally only certainty is considered. As shown recently by Gordon (1985) and Bulow and Summers (1984), the computation of effective tax rates can be affected by the precise treatment of the interactions of uncertainty and the tax code.

2. Effective tax rates and capital market equilibria

In King and Fullerton (1984) the measurement of effective tax rates is based on the computation of the tax wedge between p, the before tax marginal product on a selected stand-alone project, and s the after tax return received by a final investor. The wedge (p - s) is meant to capture all the elements of government fiscal policies which create a divergence between the returns to investment and savings. The "tax-inclusive" measure of the effective tax rate, t, is defined as this tax wedge divided by the pre-tax return[5]:

$$t = \frac{p - s}{p}.$$

The value of p is equal to the gross marginal return on a project (MRR) less the (exponential) rate of economic depreciation (δ). In equilibrium the present value of the gross marginal return net of company tax is equal to the cost of the project (set arbitrarily equal to unity) net of the present discounted value of tax allowances (A).

$$1 - A = \frac{(1 - \tau)MRR}{(\rho + \delta - \pi)}$$

The left-hand side of this expression is equal to the net cost of the project. The right-hand side is the present value of net returns where ρ is the discount rate used by the firm for its projects (or alternatively a measure of the internal rate of return), π is the rate of inflation and τ is the corporate tax rate. Solving for MRR and recalling that p = MRR - δ, we obtain that

$$p = \frac{(1 - A)}{(1 - \tau).} (\rho + \delta - \pi) - \delta$$

On the other hand, savers receive a post-tax rate of returns given by:

$$s = (1 - m)i - \pi - w_p \qquad (1)$$

where i is the nominal interest rate, w_p is the personal wealth tax and m is the marginal personal tax rate on the particular type of finance.

This framework is very general and can be applied to any form of "tax wedge" between pre and post-tax returns. The studies carried out by King and Fullerton and by others using similar approaches have considered only investments carried out within a closed economy. It is easy to see, however, that this exercise could also apply to investments which are made in countries other than those where the saver resides.

The computation of effective tax rates in the framework developed by King and Fullerton assumes that companies and individual investors operate as if there were only one form of finance at the margin. In order to "close" the model, either of two different "pseudo-equilibrium" conditions are imposed on the financial markets depending on whether profitability (fixed-p) or real interest rates ($r = i - \pi$) are considered exogenous (fixed-r).[6] Under the fixed-p case, the value of the internal rate of return on a project (ρ) is determined assuming a fixed value of the pre-tax rate of return p. The value of ρ is therefore independent of financial policy and the ownership structure of the firm. The fixed-p solution avoids problems with respect to the determination of a capital market equilibrium at the cost of entailing that a different nominal market rate of interest i prevails for each project.

As we shall see below, although the fixed-p case cannot be said to be satisfactory with respect to the nature of the capital market equilibrium condition, it helps to analyse the extent to which differences in the tax base across countries lead to implicit divergences of interest rates. Furthermore, it is possible to reconcile the fixed-p case with another pseudo-capital market equilibrium by weighting each of the values of i according to the relative importance of individual projects and sources of finance (Brennan(1971)). The overall effective tax rates in this case would depend solely on the taxation of different types of ultimate investors. As we shall show below in section 2 an alternative version of this assumption is intuitively plausible in the open economy case.

On the other hand, the fixed-r case assumes a capital market equilibrium in which the marginal yield on financial investments of different types is equalised, except for differences in tax treatment amongst owners of financial assets: "causality" runs from an exogenous, given fixed interest rate to marginal products. Here, too, the capital market equilibrium appears somewhat arbitrary, since firms face different discount rates according to the source of finance and, possibly, final ownership.[7]

The link between p and s is provided through the capital market, i.e. via the relationship between ρ and i. In the absence of taxes these two values would be equal. In general the value of ρ and i will diverge because of the different tax treatment of various sources of finance, and because the personal tax rates levied on ultimate savers may vary markedly. In order to compute a value for the effective tax rates, some authors have assumed that ρ can be determined from a weighted average of tax adjusted values of i across the various possible sources of finance and ownership. This is, however, not the only possible type of "cum-tax" equilibrium. The presence of shareholders facing different marginal tax rates and the existence of appropriate constraints can lead to equilibria in which individuals will prefer different financial policies depending on their tax characteristics. Miller (1977) and DeAngelo and Masulis (1980) have described some such segmented equilibria by constraining individuals from engaging in tax arbitrages and by assuming constraints on borrowing and short sales.

2.1 The closed economy

The approach followed by King and Fullerton, which is replicated here, is to consider separately each potential source of finance. The interrelation between ρ and i is analysed independently for each type of investor. The aggregate measures of effective tax rates are then obtained by weighting each effective tax rate by the relative importance of individual projects in the economy.

In the absence of some form of equilibrating mechanism, financial policies will be corner solutions dictated by the values of tax parameters and for a given market interest rate (i) the discount rate for valuing alternative projects (ρ) will be uniquely determined. The values of the discount rate in these circumstances are by now familiar (King (1977)). For debt finance, since nominal interest is tax deductible at a rate ϕ, which is generally equal to unity unless the firm is "tax exhausted" or there is a cash flow type corporation tax, the value of the discount rate is given by:

$$\rho = i\,(1 - \phi\,\tau) \qquad (2)$$

When financing the firm via new share issues, investors will require a net of tax dividend yield equal to the investor's opportunity cost, i.e. the rate of return on an alternative investment. The latter is given by $(1 - m_b)i$ where m_b is the marginal personal tax rate which applies to interest income. The net of tax dividend yield is $\rho\theta(1 - m_s)$, where m_s is the marginal personal tax rate on shares and θ equals the additional gross dividends shareholders could receive if one unit of post tax earnings were distributed (King, 1977).[8] Hence the firm's discount rate is given by:

$$\rho = \frac{(1 - m_b)}{\theta(1 - m_s)}\, i \qquad (3)$$

In the case of retained earnings, for a given tax rate on accrued capital gains z[9] the representative investor will require $\rho(1 - z) = i\,(1 - m_b)$ so that

$$\rho = \frac{(1 - m_b)}{(1 - z).}\, i \qquad (4)$$

Under certainty, firms will in general be pushed to finance all their investments by following at the margin the financial policy having the lowest value of ρ The choice will be dictated by the interaction between the personal, capital gains and the company tax systems. Indifference amongst alternative financial policies for the firm is achieved only for a specific set of tax parameters.[10]

2.2 The open economy

In an open economy with full capital mobility domestic savings and investment need not be equal to one another. Interest rates are determined in the world capital market where savers and investors from all countries are brought together. This section describes the characteristics of capital market equilibrum for various sources of finance in an open economy consisting for expository purposes of two countries: country j, the capital importer, and country k, the capital exporter. In the case of debt-financed investments by firms in country j by funds coming directly from country k, there would be no change as regards the discount rate ρ^j if borrowing continued to be in domestic currency and any foreign exchange gains and losses were borne by savers in country k.[11]

In the case of a purchase of new shares by foreign shareholders, the value of ρ will differ in three important respects from its domestic counterpart. Firstly, ρ^j depends on the tax treatment of foreign investors in the capital-importing country and the system of double taxation relief in the investor's home country. The net of tax dividend yield is given by $(1 - m_{sk}^{j})\ \theta_k^{j}\rho^j$. The parameter θ_k^j defines the additional dividends which shareholders would receive before personal tax if one unit of post tax earnings in country j were distributed to shareholders of country k. The value m_{sk}^j is the marginal personal tax rate in country k on dividends received by shareholders in country k after allowance for double taxation relief on foreign withholding taxes levied in country j.

The value of m_{sk}^j and θ_k^j differ by country in which the income originates, by the country of residence of shareholders and by type of final owner. Table 3 sets out the matrix of the average values of $\theta_k^j(1-m_{sk}^j)$ across different types of final investors (households, tax-emempt institutions, insurance companies) from the capital-exporting country for a select group of OECD countries. The value of θ_k^j depends on the system of integration of personal and company taxes in the capital-importing country. With a "classical system", such as that which exists in the Netherlands and the United States, the value is always unity. Under the imputation system, in most cases the value of θ_k^j equals unity because tax credits are not provided to foreign shareholders. However, under certain double taxation treaties, such as that between the United

Kingdom and France, the imputation credit is passed on to final shareholders. For Germany and Japan, where the integration between the company and personal tax system operates in part through a split-rate system, no distinction is made between domestic and foreign shareholders, and in this case too $\theta_k^j > 1$. The value of m_{sk}^j used in the computations shown in the table depends on the interaction between withholding taxes in the capital-importing country, the personal tax system in which the final beneficiaries reside, and the method by which double taxation relief is provided. If foreign income is exempt from personal tax, as it is in the case of the Netherlands, the withholding tax in the capital-importing country determines the final tax burden. On the other hand, if foreign taxes can be credited in full against personal taxes, then the value of m_{sk}^j will be given by the marginal personal tax in the capital-exporting country. It should be noted that a very important exception to the credit system applies for tax-exempt institutions which cannot claim the tax credit for foreign withholding taxes because they lack taxable income.[12] As far as possible the values shown in Table 3 take account of these complications.[13]

Foreign shareholders may suffer a capital loss (gain) on their purchases of shares owing to an exchange rate depreciation (appreciation) $\hat{e}_k^j$ ($-\hat{e}_k^j$). If this loss is allowed as an offset against capital gains tax levied at an accrual rate of z_k^j, the total return for financing through new share issues will be given by $(1 - m_{sk}^j)\theta_k^j\rho - (1 - z_k^j)\hat{e}_k^j$.

The final important difference between the closed and open economy arises with respect to the opportunity cost of alternative security investments. For foreign shareholders this value will be given by $(1 - m_{bk}^k)i^k$ where the i^k and m_{bk}^k are respectively the marginal personal tax rate on domestic interest income and the domestic interest rate in country k. The values of $1-m_{bk}^k$ for the same select group of OECD countries are shown along the principal diagonal of Table 4.

Equating the net-of-tax return on bonds in country k with the after-tax dividend yield on shares issued in country j and purchased by residents in country k, the discount rate in country j depends on the foreign rate of interest as follows:

$$\rho^j = \frac{i^k (1 - m_{bk}^k) + (1 - z_k^j)\hat{e}_k^j}{\theta_k^j(1 - m_{sk}^j)} \qquad (5)$$

Similar arguments apply in the the case of retained earnings so that the value of ρ^j will be given by:

$$\rho^j = \frac{i^k (1 - m_{bk}{}^k) + (1 - z_k{}^j)\hat{e}_k{}^j}{(1 - z_k{}^j)}. \quad (6)$$

Our discussion to this point is summarised in Table 5, which shows the value of ρ^j for different sources of finance depending on the country of residence of the final investor. As the table makes readily clear, a link is needed between domestic and foreign interest rates in order to obtain an overall equilibrium in the international financial market.

2.3 Linking domestic and foreign markets

In the absence of taxes and under certainty this link is provided by the open interest parity condition which states that the interest in country j equals the interest rates in country k plus the expected depreciation of the currency of country j.[14]

$$i^j = i^k + \hat{e}_k{}^j. \quad (7)$$

With personal taxes this equilibrium condition needs to be modified. After tax, bondholders in country k receive net interest income on investments in country j equal to

$$i^j (1 - m_{bk}{}^j) - (1 - z_k{}^j)\hat{e}_k{}^j. \quad (8)$$

Equating expression (8) with the after-tax return on domestic debt in country k $[(1 - m_{bk}{}^k)i^k]$, in the presence of taxation the equilibrium condition for debt instruments becomes:

$$i^j(1 - m_{bk}{}^j) - (1 - z_k{}^j)\hat{e}_k{}^j = i^k(1 - m_{bk}{}^k) \quad (9)$$

As can be seen from (9), open interest parity (expression (7)) occurs only if the same marginal tax rate applies to both types of interest payments and to foreign exchange gains (and losses). If $m_{bk}{}^k = m_{bk}{}^j (\neq z_k{}^j)$ this condition is reminiscent of the Fisher Hypothesis modified for taxes as suggested by Darby (1976), Feldstein (1976) and Tanzi (1976).[15] Indeed, if purchasing power parity holds, whereby changes in nominal exchange rates are determined by inflation differentials ($\hat{e}_k{}^j = \pi^j - \pi^k$), and if world interest rates and inflation are given:

$$\frac{\delta i^j}{\delta \pi^j} = \frac{1 - z_k^{\ j}}{1 - m_{bk}^{\ k}}.$$

Unfortunately, the relationship between domestic and foreign interest rates is more complicated and depends in a crucial manner on the interaction between four sets of factors:

- the entities carrying out the interest rate arbitrage and limitations which are set thereon;
- whether the country is small in the sense that it takes interest rates as given abroad;
- the tax treatment of capital gains on foreign exchange rate gains and losses;
- the specific type of interaction between interest rates, inflation, exchange rates and taxes.

(a) Two sets of general complications arise when considering the taxable entity operating at the margin and determining the implicit value of ρ Firstly, different relative valuations with respect to coupon payments and capital gains by investors in various countries may give rise to asymmetries and "abnormal" two-way capital movements. In the case where marginal personal taxes in the two countries are equal for domestic and foreign investments, this may result in interest parity not being achieved if the following two conditions occurred:

$$(1 - m_{bj}^{\ j})(i^j - i^k) > (1 - z_j^{\ k})\hat{e}_k^{\ j} \tag{10a}$$

$$(1 - m_{bk}^{\ k})(i^j - i^k) < (1 - z_k^{\ j})\hat{e}_k^{\ j}. \tag{10b}$$

For example, investors with low $m_{bj}^{\ j}$ and high $z_j^{\ k}$ would prefer investing domestically, whereas investors with high $m_{bk}^{\ k}$ and low $z_k^{\ j}$ would prefer investing abroad. In this situation there would be no determinable tax equilibrium for certain values of $\hat{e}_k^{\ j}$.[16] This possibility was first noted by Levi (1977). Its full implications in a valuation model of international assets have been recently examined by Gordon (1987) who draws the analogy between a multi-country framework and the reasoning which applies to the determination of equilibrium interest rates in a closed economy in the presence of a progressive marginal tax rate with different treatments of various financial assets. In these circumstances, the equilibrium which emerges is one in which investors are segmented and specialise in different tax (and country) preferred assets. Indeed, unless the tax parameters and exchange rate changes are correctly aligned, Gordon concludes that foreign securities are generally the tax preferred asset.

The second set of elements affecting the determination of the international equilibrium concerns the group of domestic or foreign investors carrying out the arbitrage between domestic and foreign interest rates across countries. Expression (8) assumes implicitly that this is done by individual investors; however, in all likelihood this arbitrage will be carried out by companies taxed at a rate τ which differs from that on interest earnings of the personal sector. Furthermore, even amongst individual investors the values of m and z may differ widely. For example, in the United States and the United Kingdom pension funds are not taxed domestically and are therefore very responsive to the level of foreign withholding taxes.

(b) In expression (7) the value of i^k - which can be interpreted as the world interest rate - is assumed as fixed and, given a value of $\hat{e}_k{}^j$, it is possible to determine i^j. However, this small country assumption is unsatisfactory for most of the countries being considered in this paper because of their size and because of the considerable interaction between their financial markets.

Intuitively, the value of the world interest rate could be determined as an average of domestic interest rates $(\Sigma_j \alpha_j i^j)$, weighted by the relative importance (α_j) of individual countries in international financial markets. Under conditions of uncertainty Solnik (1973) derived such weights explicitly for the international equivalent of the capital asset pricing model. In his model world interest rates depended on national wealth, net investment abroad, the co-variance of exchange rates and the degree of risk tolerance of investors. Because of open interest parity and complete certainty, in Section 3 of this paper we have neglected risk aversion and the co-variance of exchange rates in modelling world interest rates. Two alternative weighting schemes have been used as proxies for national wealth and the foreign investment position. GNP has been used as the closest approximation to the relative value national wealth, whereas gross foreign investments of the private sector have been taken as a mixed measure of wealth and foreign presence.[17]

(c) As shown in Table 6, the tax treatment of foreign exchange gains and losses differs markedly across countries and is very complex because it often relies heavily on case law. There is also no evidence of the revenue implications of these provisions. Therefore in the empirical section of this paper we have abstained from calculating any value for $z_k{}^j$ from existing tax laws.

In addition to these implications for capital market equilibrium, the tax treatment of foreign exchange gains and losses enters into the returns to savers in country k (s_k^j) unless purchasing power parity adjustments are made. Accordingly, expression (1) becomes:

$$s_k^j = (1 - m_k^j)i^j + (1 - z_k^j)\hat{e}_k^j - w_k^j$$

or

$$s_k^j = (1 - m_k^j)i^j - \hat{e}_k^j - w_k^j$$

where w_k^j is the wealth tax in country k on investments in country j.

3. Empirical results

Using the methods described above, we now turn to various empirical simulations which were carried out to assess the impact of capital mobility on effective tax rates for the principal EC countries. The study is limited to the effective tax rates within each single country as in King and Fullerton, and not for the full range of possible combinations of countries and investors. Effective tax rates are computed for the same 81 investments as in King and Fullerton with the major effect of "openness" being that of altering the value of interest rates faced by domestic savers and investors. Hence no measure of the effective tax rates on cross-border investments is provided.[18]

Before turning to the open economy it is useful to review briefly the value of effective tax rates for the closed economy under both the fixed-p and fixed-r cases. Table 7 reports the value of effective tax rates for the fixed-p case[19] at end-1986 under the assumptions of zero and 10 per cent inflation. As can be seen from the bottom line in the top and bottom panels, there were pronounced differences in the overall value of the effective tax rates across countries. Most of this variation does not result from differences in company tax rates or system of integration between the corporate and personal taxes. Rather, the differences are accounted for by varying types depreciation allowances, regional investment incentives, and other levies such as wealth taxes.[20]

As far as country rankings are concerned, at zero inflation the Italian tax system displays the lowest rates because of the generous system of depreciation allowances and investment incentives but also because of the low personal tax rates on interest income. By contrast, the French tax system is particularly onerous in view of high marginal personal tax rates. Inflation tends to widen the range of effective tax rates between the various countries (particularly between France and Italy) and across assets. Differences in effective tax rates on particular assets because of inflation result from the extent to which tax provisions allow implicitly for some form of adjustment against price movements. For example, the taxation of inventories is unaffected by inflation in countries having LIFO valuation methods. Another interesting finding is that in the case of Germany the overall effective tax rate declines with inflation. This is due essentially to the deductibility of interest payments and of wealth taxes on debt finance, the benefit of which increases markedly with inflation.[21] Since debt finance accounts for a sizeable share of German firms' sources of finance these effects overwhelm the negative impact recorded for other sources of finance. In the case of the United Kingdom, which until 1984 was a low tax country because of the generous depreciation provisions and stock allowances, inflation increases effective tax rates quite markedly.

Table 8 reports similar computations for the fixed-r case. The numerical values of the effective tax rates differ from those shown in Table 7 although qualitatively the ranking of countries remains unchanged. In particular, under the fixed-r case in several instances the effective tax rates take on extreme values such as the case of Italy and the Netherlands for investments in machinery. In several cases not shown in this table these values were even more extreme for certain projects financed via debt.[22]

It is useful to examine the open economy implications for the fixed-p and fixed-r cases separately since each of these type of simulations can be used to answer different sets of questions in the presence of capital mobility. However, several common assumptions to both sets of simulations should be mentioned. Firstly, for lack of better information exchange rate gains and losses are assumed to be untaxed ($z_k^j = 0$). Secondly, it should be noted that in all the simulations we have assumed that US and Japanese investors take part in the European capital market and therefore affect the average values of interest rates. In the fixed p-case, however, these effects come only from the tax treatment of savings and do not take account of the effect on world interest rates coming from the tax treatment of companies in these two countries.[23] Finally, we have allowed for the foreign influence on domestic interest rates to occur only through the average values of taxes on cross-border transactions weighted by GNP as computed in Tables 3 and 4.

3.1 Capital mobility and effective tax rates: the fixed-p case

The fixed-p case assumes a uniform value of p across countries from which the value of the internal rate of return ρ is derived independently of financial markets and the degree of capital mobility. Graph 1 shows the different values of ρ for the EC countries against domestic rates of inflation. In the fixed-p case these lines can be interpreted as the highest cost of capital which firms are willing to pay for a given real marginal product. As can be seen from the graph the values of ρ are highest for Italian firms and rise by roughly 0.9 per cent for every percentage point increase in the rate of inflation. Similar behaviour with respect to changes in the rate of inflation occurs for Belgium and the Netherlands albeit at lower absolute values of ρ For high taxed countries having the lowest values of ρ the ranking changes with the rate of inflation. At low rates of inflation France has a higher value of ρ than Germany; however, as inflation rises, for Germany, ρ increases by nearly 0.9 for every percentage point change in inflation, whereas for France the value of ρ goes up comparatively slowly (only by .55 per cent.). These differing values of ρ in the fixed-p case which apply to both the closed and open economy provide us with the potential range of variation in interest rates coming from differences in tax systems applying to corporate income before firms face the financial markets.

In order to derive the values of effective tax rates in the open economy we need to take account of the constraints imposed by the existence of foreign shareholders and by the exchange rate on the value of i^j. For equity finance, the values of i^j in each country can be computed in a relatively straightforward fashion by inverting the values of expression (5) and (6), and by employing the average values of $(1 - m_{bk}{}^j)$ and $\theta(1 - m_{sk}{}^j)$ shown in Tables 3 and 4. In the case of debt finance, however, the existence of foreign investors does not affect the value of i^j which is determined solely by the tax rate on corporate profits (expression 2). The computed values of i^j then enter into the returns to savings for domestic shareholders shown in expression (1). Thus, the impact on effective tax rates of moving to an open economy with full capital mobility results solely from the differences in the tax rates on equity finance for domestic investors and the weighted average of international investors.

Table 9 reports the <u>changes</u> in effective tax rates resulting from the assumption of an open economy. As can be seen from the table in the case of zero inflation the most significant effect of assuming capital mobility is that the range of differences across countries narrows significantly. This result is what would be expected and is implicitly built into the model since the confluence of investors into a unique market tends to reduce relative variations of alternative financial policies across countries.

Although Italy remains the country with the lowest overall effective tax rate, on average the rates increase by 19 per cent. and in the case of retained earnings the increase is even larger (43 per cent.). At the other extreme, Germany actually displays a reduction in marginal effective tax rates. The reason for the differences between the closed and open economy simulations result from the after tax relative valuations placed by domestic and international shareholders on streams of income, i.e. $m_{bk}{}^j$, $m_{sk}{}^j$ and $\theta_k{}^j$ are not equal to their domestic equivalents. In a closed economy these relative valuations are determined by the rates of tax on personal income in individual countries. In an open economy they depend in an intricate fashion on withholding taxes in the capital importing country, on the rate of personal income taxes in the capital exporting countries, and on the relative weight of each economy in the world. Moreover, these differences are not simply the result of aggregating marginal personal tax rates across countries since the tax treatment of foreign source income may differ very markedly from that which applies to domestic income. It is, therefore, not possible to make a one-for-one correspondence between the burden of domestic taxes appropriately weighted by GNP and that which applies to foreign investment.

Turning to the lower panel of Table 9, with domestic and world inflation rates of 10 per cent. in most instances the convergence of effective tax rates is even more striking. However, in the case of Germany, capital mobility has the effect of raising nominal interest rates received by final shareholders above the level which would exist in the absence of capital mobility and thereby to reduce effective tax rates.

3.2 Capital mobility and effective tax rates: the fixed-r case

In the fixed-r case it is possible to allow for a greater number of effects deriving from capital mobility on the determination of domestic interest rates than the fixed-p case is capable of capturing. Firstly, interest rates across countries are linked through open interest parity (expression (9)). It is therefore possible to consider the effect of openness on all forms of finance. Secondly, it is possible to derive a single world nominal interest rate i* given by the weighted average of domestic inflation rates adjusted by the weighted average of marginal personal tax rates as follows:

$$i^* = \bar{r} + \frac{\Sigma\alpha_j\pi^j}{1 - \Sigma\alpha_j m_{bk}{}^j}$$

where $\Sigma\alpha_j m_{bk}{}^j$ refers to the worldwide GNP-weighted average personal tax rate on bonds[24] and r (= .05) is the after-tax real interest rate. The values of nominal interest rates in individual countries are linked to this world interest rate according to expression (9). Finally, it is possible to examine the impact of having different rates of inflation across countries on the measurement of effective tax rates.

Table 10 displays the difference between the simulations shown in Table 8 and the new simulations carried out assuming capital mobility. With zero inflation the findings confirm our previous results that capital mobility brings about convergence of effective tax rates across countries and at the same time raises their value. The lower panel for an inflation rate of 10 per cent. at home and abroad displays somewhat different findings from the fixed-p case. These differences result to a large extent from the extreme values taken in the case of Belgium, Germany and Italy by some debt financed investments. As already noted in King and Fullerton this occurs because the advantages of debt finance are so great that the revenue generated by a project need not cover even depreciation costs to produce the stimulated rate of return. Very low and possibly negative real rates of return (net of depreciation) may be consistent with equilibrium in the capital market with investors earning positive real returns on their savings.

A special set of simulations were carried out for the fixed-r case to examine the effects of having significant inflation differentials between the domestic economy and the "rest of the world". The results shown in Table 11 suggest that the rate of domestic inflation is the main determinant of the rate of effective tax rates for these countries.

4. Conclusions

During the 1960s, much attention was paid to devising systems of commodity taxes which would be non-distortionary with respect to trade, which at the time was expanding very rapidly. The EC countries went further and established a significant degree of harmonisation by adopting a value added tax based on a destination principle. The growth of capital movements and deregulation of financial markets in the eighties is of equal importance to the opening up of trade during the postwar period and raises similar questions for tax policy.

Most of the discussions regarding international comparisons of tax burdens on savings and investment have focused attention on the taxation of corporate entities and on the personal tax treatment of savings. These are indeed the areas in which it is likely that agreement across countries can be easily reached with respect to some broad form of tax harmonisation. They are also the most visible types of taxes. It is a merit of the King-Fullerton study to have extended the cross country comparison to include local taxes as well as wealth taxes. However, even that study has assumed a closed economy and this might have seriously biased upward the actual extent of cross-country differences. Indeed, King and Fullerton concluded that there were major differences in effective tax rates across countries.[25] This paper extends their findings to an open economy in an attempt to examine some of the possible implications of operating widely divergent tax systems within a highly integrated capital market.

The principal conclusion that can be derived from the simulations carried out in section 2 is that the existence of an international capital market tends to reduce differences in effective tax rates across countries resulting from the tax treatment of alternative types of savings and investment. Differences, however, do remain and may be the source of competitive inequalities.

Another finding which needs to be examined more closely is that capital mobility, under the existing tax systems and types of double taxation agreements, tends to raise effective tax rates. This result is probably a reflection of our literal interpretation of existing tax provisions rather than of the reality of the tax treatment of international flows of funds. Individual investors often shelter their income from foreign investments in tax havens with a favourable tax climate or undertake transactions which transform a high taxed stream of income into another having a lower rate. These factors cannot be captured in our model.

It is clear that as we develop our model, it will be possible to subject the numerous other questions. In particular, it should be possible to examine the impact on our results of assuming varying degrees of openness to external shocks. In particular it should be possible to report the effects of tax reforms such as those recently carried out in the United States on effective tax rates in Europe and to examine some of the hypotheses put forward in Sinn (1985).

FOOTNOTES

* This paper has appeared in the Finnish Economic Papers and is reprinted with permission from the Finnish Society for Economic Research. We appreciate the comments on a previous version of this paper from several colleagues and from participants of the Helsinki Conference on "Perspectives on Capital Income Taxation in Europe". Paul Van den Bergh kindly supplied us with the figures shown in Tables 1 and 2. The views expressed in this paper are our own and not necessarily those of the Bank for International Settlements.

1 There are three possible arguments against the finding of Feldstein and Horioka. Firstly, if the rate of time preference and technology are sufficiently similar across countries, domestic savings and investment would be brought into equality even in a perfectly integrated world [Stulz (1986)]. Secondly, for most of the sample period used by Feldstein and Horioka the major countries operated with fixed exchange rates. Under the "rules of the game" of the fixed exchange rate system countries followed macro-economic policies aimed at maintaining current-account balance, thus leading to equalisation of saving and investment. The "rules of the game" have not changed markedly since the inception of floating rates and this may help to explain the continuing balance between saving and investment which has been reported for more recent years. Finally, common cross-country exogenous shocks may affect savings and investments in a similar fashion [Summers (1985)].

2 The effective tax rates computed in this paper, like those commonly produced in the literature, assume unchanging tax rules and do not take account of expectational factors or adjustment costs.

3 We do not, however, consider the manner in which the specific tax treatment of banks might affect cross-border lending decisions (see Frankel (1984) and Alworth (1984)).

4 See Slemrod (1987) for a survey of these issues.

5 In their study King and Fullerton compile effective tax rates on 81 different projects broken down by asset, industry, source of finance and ownership. Overall country-wide effective tax rates are averages weighted by the estimated importance of each of these projects in total capital stock.

6 An alternative procedure would be to specify a general equilibrium model having production technologies and intertemporal utility functions.

7 See Scott (1987) for critique of the fixed-r equilibrium and an alternative procedure.

8 The value of θ will be greater than unity if there is some form of integration between company and personal taxes. For example, under the imputation system the tax credit received by the shareholder is equal to $(\theta - 1)$.

9 See King and Fullerton (1984) for a discussion on how to transform realised into accrued tax rates.

10 The nature of market equilibrium with respect to the financial policies followed by the firm and the specific value taken by the cost of capital depends on two sets of factors: the tax laws and legal constraints on arbitrages. In practice, these two sets of factors interact because tax laws not only impose corner solutions with respect to financial policies but also encourage arbitrages amongst similar streams of income having a differential tax treatment. Legal constraints are necessary to prevent firms from fully exploiting tax avoidance strategies (King, 1977).

11 However, domestic interest rates are linked to world interest rates via interest parity as shown below.

12 This also applies to those shareholders whose marginal personal tax rate is less than the foreign withholding tax.

13 See Alworth (1987b), Chapter 4, for a more complete description of provision concerning double taxation relief in various countries.

14 Under certainty and with perfect foresight open interest parity is equivalent to covered interest parity which states that $i^j = i^k + f$, where f is the forward discount on the exchange rate. In the presence of uncertainty

and risk aversion, covered and open interest parity may differ on account of potential risk premia.

15 Assuming the average real return (r) is constant across inflation rates so that the average saver receives $s = i(1 - m) - \pi$, the nominal interest rate must equal $i = r + \pi/(1 - m)$.

16 To put this conclusion somewhat differently, if domestic interest rates are determined by the behaviour of foreign investors (expression 10b), for specific constellations of interest and exchange rates domestic investors would have incentives to borrow without limit in the international capital market to buy up domestic bonds. See Sorensen (1986) for similar arguments.

17 The values of the weights employed in the simulations of Section 3 are available on request from the authors.

18 For such estimates in the case of direct investment see Alworth (1987b).

19 p is assumed to equal 10 per cent.

20 See Alworth (1987a).

21 This result contrasts markedly with that reported in King and Fullerton. The additional benefits from leverage due to the deductability of borrowings from wealth tax assessments was not considered in that study.

22 See section 2.2 for a possible explanation of this result.

23 It would, for example, not be possible to examine the impact of US tax changes with respect to depreciation allowances on interest rates in Europe. See Sinn (1985) for some conjectural exercises along these lines.

24 The value of $\bar{r}$, the real after tax interest rate, is taken as fixed and exogenous.

25 Another problem with assessing the results of King and Fullerton resides in other differences across countries. Some result from the interaction of the macro-economy with tax rules which are not fully captured in King and Fullerton, such as the phenomenon of tax exhaustion; others, from differences in the character of the provision of public goods.

REFERENCES

Alworth, J.S. (1984): "The regulatory and accounting environment in six European countries for the private financing of LDCs", mimeo.

Alworth, J.S. (1987a): "Taxation and the cost of capital in six EC countries." in Cnossen, S. ed. Tax Coordination in the European Community, Deventer: Kluwer.

Alworth, J.S. (1987b): Finance and Investment Decisions of Multinational Companies: The Tax Aspects Oxford: Blackwells

Bradford, D. and Fullerton, D. (1981): "Pitfalls in the construction of effective tax rates." in Hulten, C.R. (ed.) Depreciation, inflation and the taxation of income from capital. Washington: The Urban Institute

Brennan, M.J. (1970): "Taxes, market valuation and corporate financial policy." National Tax Journal, 23, pp. 414-27

Bulow, J. and Summers, L. (1984): "The taxation of risky assets." Journal of Political Economy, 92, pp. 20-39

Darby, M. (1975): "The financial and tax effects of monetary policy on interest rates." Economic Inquiry, 13, pp.266-74

Feldstein, M. (1976): "Inflation, income taxes and the rate of interest: a theoretical analysis." American Economic Review, 66, pp.809-20

Feldstein, M. and Horioka, C. (1980): "Domestic savings and international capital flows." The Economic Journal, 90, pp.314-29

Frankel, A. (1984): "Federal taxation and the domestic-foreign asset choice of a US bank." Federal Reserve Board Discussion Paper

Gordon, R. (1985): "Taxation of corporate capital income: tax revenues versus tax distortions." Quarterly Journal of Economics, 100, pp. 1-27

Gordon, R. (1986): "Taxation of investment and savings in a world economy." American Economic Review, vol.76, pp. 1086-1101

King, M. and Fullerton, D. eds. (1984): The Taxation of Income from Capital, Chicago: The University Press
Levi, M. (1977): "Taxation and 'abnormal' capital flows." Journal of Political Economy, 85, pp.635-46
Obstfeld, M (1986a): "Capital mobility in the world economy: theory and measurement." in The National Bureau Method, International Capital Mobility and Other Essays, Ed. by Brunner and Meltzer, Carnegie Rochester Conference Series on Public Policy
Obstfeld, M. (1986b):"How integrated are world capital markets? Some new tests." NBER Working Papers. no.2075
Scott, M. (1987): "A note on King and Fullerton' formulae to estimate the taxation of income from capital.", mimeo
Sinn, H. (1985): Kapitaleinkommenbesteurung, Tubingen: Mohr
Sinn, H. (1985): "Why taxes matter: Reagan' accelerated cost recovery system and the US trade deficit." Economic Policy, 1, pp. 239-46
Slemrod, J. (1987): "International capital mobility and the theory of capital income taxation.", mimeo
Solnik, B. (1973): European Capital Markets, Lexington: Lexington Books
Sorensen, P.B. (1986): "Taxation, inflation and asset accumulation in a small open economy." European Economic Review, 30, pp.1025-41
Stulz, R. (1986): "Comment capital mobility in the world economy: theory and measurement." in The National Bureau Method, International Capital Mobility and Other Essays, Ed. by Brunner and Meltzer, Carnegie Rochester Conference Series on Public Policy
Summers, L.H. (1985): "Taxation and international competitiveness", mimeo
Tanzi, V. (1976): "Inflation, indexation and interest income taxation." Banca Nazionale del Lavoro Quarterly Review, 116, pp.64-76
Tanzi, V. (1984): ed. Taxation, Inflation and Interest Rates, Washington D.C.: International Monetary Fund

Table 1

Cross-country correlations of changes in money market interest rates

Period: 1971-74

	US	CA	JP	GB	DE	FR	IT	NL	BE
US	1.00	0.51	0.15	0.19	0.19	0.22	0.32	0.36	0.39
CA		1.00	0.17	- 0.23	0.18	0.02	0.24	0.41	0.27
JP			1.00	0.12	0.05	0.32	0.21	0.26	0.29
GB				1.00	0.07	0.14	0.13	0.19	0.12
DE					1.00	0.23	- 0.03	0.36	0.06
FR						1.00	0.31	0.38	0.36
IT							1.00	0.13	0.59
NL								1.00	0.29
BE									1.00

Period: 1975-79

	US	CA	JP	GB	DE	FR	IT	NL	BE
US	1.00	0.64	0.19	0.24	0.42	0.22	0.19	0.07	0.24
CA		1.00	0.07	0.06	0.23	0.08	0.37	0.10	0.30
JP			1.00	0.04	0.24	0.43	0.23	0.23	0.20
GB				1.00	0.34	0.28	0.10	0.10	0.26
DE					1.00	0.52	0.21	0.27	0.36
FR						1.00	0.26	0.28	0.48
IT							1.00	0.07	0.39
NL								1.00	0.49
BE									1.00

Period: 1980-86

	US	CA	JP	GB	DE	FR	IT	NL	BE
US	1.00	0.72	0.26	0.13	0.09	0.25	- 0.01	0.13	0.28
CA		1.00	0.15	0.30	0.29	0.26	0.15	0.29	0.35
JP			1.00	0.06	0.18	0.24	0.01	0.04	0.25
GB				1.00	- 0.03	0.05	- 0.02	0.08	0.24
DE					1.00	0.22	0.21	0.48	0.55
FR						1.00	0.30	0.23	0.17
IT							1.00	0.04	0.24
NL								1.00	0.10
BE									1.00

Table 2

Cross-country correlations of changes in bond yields

Period: 1971-74

	US	CA	JP	GB	DE	FR	IT	NL	BE
US	1.00	0.61	0.04	0.13	0.26	0.10	- 0.02	0.26	0.29
CA		1.00	0.06	0.17	0.30	0.25	0.13	0.29	0.27
JP			1.00	0.08	0.04	0.32	0.02	0.34	0.28
GB				1.00	0.03	0.30	0.21	- 0.10	0.17
DE					1.00	0.16	0.24	0.39	0.08
FR						1.00	0.27	0.20	0.35
IT							1.00	- 0.05	- 0.06
NL								1.00	0.28
BE									1.00

Period: 1975-79

	US	CA	JP	GB	DE	FR	IT	NL	BE
US	1.00	0.56	0.24	0.17	0.26	- 0.23	0.07	0.22	0.32
CA		1.00	0.10	0.24	0.18	0.10	0.23	0.23	0.15
JP			1.00	0.06	0.48	0.06	0.12	0.20	0.34
GB				1.00	0.28	0.29	0.36	- 0.03	0.15
DE					1.00	0.17	0.18	0.45	0.29
FR						1.00	0.28	- 0.03	0.19
IT							1.00	0.00	0.14
NL								1.00	0.02
BE									1.00

Period: 1980-86

	US	CA	JP	GB	DE	FR	IT	NL	BE
US	1.00	0.72	0.50	0.42	0.56	0.37	0.03	0.57	0.39
CA		1.00	0.34	0.41	0.36	0.21	0.10	0.36	0.30
JP			1.00	0.27	0.45	0.31	- 0.08	0.46	0.11
GB				1.00	0.18	0.20	0.08	0.13	0.17
DE					1.00	0.46	0.15	0.71	0.38
FR						1.00	0.28	0.47	0.46
IT							1.00	0.06	0.20
NL								1.00	0.34
BE									1.00

Table 3

Average value of $\theta_k^j(1-m_{sk}^j)$ for dividend payments

(for domestic shareholders average value of $\theta_j^j(1-m_{sj}^j)$ taken)

Capital exporting country (k)	Capital importing country (j)							
	Belgium	France	Germany	Italy	Japan	Netherlands	United Kingdom	United States
Belgium	**0.750**[1]	0.990[2]	1.932	0.660	0.750	0.660	0.720	0.660
France	0.681	**0.912**	1.571	0.681	0.774	0.681	0.962[2]	0.681
Germany	0.603	0.911[2]	**1.280**	0.559	0.686	0.603	0.641	0.603
Italy	0.672	0.672	1.420	**0.919**	0.764	0.732	0.732	0.672
Japan	0.748	0.748	1.700	0.748	**0.759**	0.748	1.042[2]	0.748
Netherlands	0.850	1.275[2]	1.930	0.676	0.966	**0.688**	1.233[2]	0.850
United Kingdom	0.715	1.073[2]	1.626	0.663	0.813	0.715	**1.064**	0.715
United States[3]	0.608	0.913[2]	1.383	0.608	0.692	0.608	0.851[2]	**0.644**
Geometric mean of $\theta^*(1-m_s^*)$								
weighted by external assets	0.675	0.906	1.524	0.658	0.746	0.669	0.905	0.685
weighted by GNP	0.651	0.883	1.469	0.650	0.724	0.650	0.878	0.670

1. Assumes all shareholders choose précompte liberatoire.
2. Shareholders receiving "avoir fiscale" or tax credit.
3. Before US tax refore of 1986.

Note: Weighted average for individual cell assumes that the ownership composition of foreign portfolios is the same as that for domestic investments. If the withholding tax in the capital importing country is greater than the marginal personal tax rate, the withholding tax is considered as final payment.

Source: Coopers and Lybrand (1986) International Tax Summaries.

Table 4

Average value of $(1-m_{bk}^{j})$ for interest payments

(for domestic interest payments average value of $(1-m_{bj}^{j})$ taken)

Capital exporting country (k)	Capital importing country (j)							
	Belgium	France	Germany	Italy	Japan	Netherlands	United Kingdom	United States
Belgium	**0.769**	0.776	0.791	0.769	0.769	0.791	0.769	0.791
France	0.582	**0.603**	0.603	0.582	0.589	0.603	0.589	0.603
Germany	0.715	0.780	**0.780**	0.780	0.738	0.780	0.780	0.780
Italy	0.775	0.783	0.798	**0.760**	0.783	0.798	0.700	0.798
Japan	0.767	0.775	0.790	0.767	**0.790**	0.790	1.775	0.790
Netherlands	0.706	0.723	0.757	0.706	0.723	**0.757**	0.757	0.757
United Kingdom	0.726	0.733	0.747	0.726	0.733	0.747	**0.747**	0.747
United States	0.768	0.804	0.804	0.768	0.780	0.804	0.804	**0.804**
Geometric mean of $(1-m_{b}^{*})$ weighted by external								
assets	0.742	0.766	0.775	0.750	0.756	0.775	0.766	0.775
weighted by GNP	0.745	0.773	0.779	0.750	0.760	0.779	0.769	0.779

Note: Weighted average for individual cell assumes that the ownership composition of foreign portfolios is the same as that for domestic investments. If the withholding tax in the capital importing country is greater than the marginal personal tax rate, the withholding tax is considered as final payment.

Source: Coopers and Lybrand (1986) International Tax Summaries.

Table 5

Marginal cost of finance (&rho.) in a closed and open economy

	Closed economy	Open economy
Debt	$i^j (1 - \varphi\tau)$	$i^j (1 - \varphi\tau)$
New share issues	$i^j \dfrac{(1 - m_{bj}{}^j)}{\Theta_j{}^j(1 - m_{sj}{}^j)}$	$\dfrac{i^k (1 - m_{bk}{}^j) + (1 - z_k{}^j)\, \hat{e}_k{}^j}{\Theta_k{}^j(1 - m_{sk}{}^j)}$
Retained earnings	$i^j \dfrac{(1 - m_{bj}{}^j)}{(1 - z_j{}^j)}$	$i^k \dfrac{(1 - m_{bk}{}^j) + (1 - z_k{}^j)\, \hat{e}_k{}^j}{(1 - z_k{}^j)}$
memo: "open" interest parity		$i^j = \dfrac{i^k (1 - m_{bk}{}^j) + (1 - z_k{}^j)\, \hat{e}_k{}^j}{(1 - m_{bk}{}^j)}$

Table 6

Taxation of foreign exchange gains and losses in selected countries

	Assets		Liabilities	
	Short term	Long term	Short term	Long term
Belgium	I-R[1]	I-R[1]	I-R[1]	I-R[1]
Canada	I-A[2]	C-R	I-A[2]	C-R
France	I-A	I-A	I-A	I-A
Germany	I-R[1]	I-R[1]	I-R[1]	I-R[1]
Italy	I-R	I-R	I-R	I-R
Japan	I-R	I-R	I-R	I-R
Netherlands ...	I-R[1]	I-R[1]	I-R[1]	I-R[1]
United Kingdom	I-R	E[3]	I-R	E
United States..	I-R	C-R	I-R	C-R

Note: A - treated on accrual basis; R - treated on realisation; I - taxed or relieved as ordinary income; C - taxed or relieved as capital gains; E - excluded from tax treatment. Definitions of short term or long term may differ. In Canada, for example, the distinction is based on whether the gain or loss on foreign exchange was associated with a purchase or sale of a capital asset.

1. Unrealised losses deductible; realised gains may be deferred.
2. The tax authorities permit the reporting of short-term gains and losses on a realisation basis. The tax treatment must be consistent over time.
3. Taxable as capital gains if equity, "debt on a security", bank deposit (Capital Gains Tax Act 1979 s. 135) or cash.

Source: Modi (1980), Chown (1986) and Peat, Marwick and Mitchell (1979).

Table 8

Effective tax rates under the existing tax system (end-1986)
fixed-r case

0 per cent. inflation

	Belgium	France	Germany	Italy	Nether-lands	United Kingdom
Asset						
Machinery	-12	21	52	-413	-176	13
Buildings	14	59	55	-48	33	52
Inventories	43	59	65	37	48	39
Industry						
Manufacturing	17	46	57	-37	10	29
Other industry	13	35	61	-50	5	38
Commerce	24	52	57	-24	26	46
Source of finance						
Debt	-12	37	14	-103	-21	23
New share issues	52	58	69	24	53	33
Retained earnings	28	47	73	-13	26	39
Owner						
Households	23	67	69	-31	39	50
Tax-exempt institutions	14	29	33	-30	-51	19
Insurance companies	1	21	30	-77	-20	35
Overall	18	47	58	-36	12	36

10 per cent. inflation

	Belgium	France	Germany	Italy	Nether-lands	United Kingdom
Asset						
Machinery	47	83	69	-17	20	67
Buildings	39	85	60	-10	48	74
Inventories	33	81	70	25	34	68
Industry						
Manufacturing	41	83	69	1	34	68
Other industry	41	83	73	- 3	35	72
Commerce	43	84	62	1	34	72
Source of finance						
Debt	-70	80	-238	-411	-100	62
New share issues	85	88	91	79	83	69
Retained earnings	66	83	94	52	60	71
Owner						
Households	59	102	94	18	89	95
Tax-exempt institutions	29	69	9	12	-123	39
Insurance companies	-8	60	-31	-129	-28	67
Overall	41	83	69	0	34	70

Table 8

Effective tax rates under the existing tax system (end-1986)
fixed-r case
0 per cent. inflation

	Belgium	France	Germany	Italy	Nether-lands	United Kingdom
Asset						
Machinery	-12	21	52	-413	-176	13
Buildings	14	59	55	-48	33	52
Inventories	43	59	65	37	48	39
Industry						
Manufacturing	17	46	57	-37	10	29
Other industry	13	35	61	-50	5	38
Commerce	24	52	57	-24	26	46
Source of finance						
Debt	-12	37	14	-103	-21	23
New share issues	52	58	69	24	53	33
Retained earnings	28	47	73	-13	26	39
Owner						
Households	23	67	69	-31	39	50
Tax-exempt institutions	14	29	33	-30	-51	19
Insurance companies	1	21	30	-77	-20	35
Overall	18	47	58	-36	12	36

10 per cent. inflation

	Belgium	France	Germany	Italy	Nether-lands	United Kingdom
Asset						
Machinery	47	83	69	-17	20	67
Buildings	39	85	60	-10	48	74
Inventories	33	81	70	25	34	68
Industry						
Manufacturing	41	83	69	1	34	68
Other industry	41	83	73	- 3	35	72
Commerce	43	84	62	1	34	72
Source of finance						
Debt	-70	80	-238	-411	-100	62
New share issues	85	88	91	79	83	69
Retained earnings	66	83	94	52	60	71
Owner						
Households	59	102	94	18	89	95
Tax-exempt institutions	29	69	9	12	-123	39
Insurance companies	-8	60	-31	-129	-28	67
Overall	41	83	69	0	34	70

Table 9

Effective tax rates assuming full capital mobility minus effective tax rates assuming no capital mobility fixed-p case

0 per cent. domestic inflation and 0 per cent. world inflation

	Belgium	France	Germany	Italy	Nether-lands	United Kingdom
Source of finance						
Debt	0	0	0	0	0	0
New share issues ...	6	- 9	- 7	10	2	2
Retained earnings ..	27	38	- 9	43	30	2
Overall	16	13	- 5	19	15	2

10 per cent. domestic inflation and 10 per cent. world inflation

	Belgium	France	Germany	Italy	Nether-lands	United Kingdom
Source of finance						
Debt	0	0	0	0	0	0
New share issues ...	26	11	3	29	14	12
Retained earnings ..	60	40	-29	69	63	7
Overall	33	24	-12	43	33	7

Table 10

Effective tax rates assuming full capital mobility minus effective tax rates assuming no capital mobility fixed-p case

0 per cent. domestic inflation and 0 per cent. world inflation

	Belgium	France	Germany	Italy	Nether-lands	United Kingdom
Source of finance						
Debt	0	0	0	0	0	0
New share issues ...	11	4	- 1	26	9	22
Retained earnings ..	10	9	4	24	8	-10
Overall	9	7	4	22	8	- 5

10 per cent. domestic inflation and 10 per cent. world inflation

	Belgium	France	Germany	Italy	Nether-lands	United Kingdom
Source of finance						
Debt	-	19	-	-	-78	1
New share issues ...	10	10	14	17	17	5
Retained earnings ..	21	14	8	36	13	13
Overall	-24	14	16	45	17	- 2

- = undefined.

Table 11

Effective tax rates assuming full capital mobility minus
effective tax rates assuming no capital mobility
fixed-p case

0 per cent. domestic inflation and 10 per cent. world inflation

	Belgium	France	Germany	Italy	Nether-lands	United Kingdom
Source of finance						
Debt	- 3	0	0	-13	1	0
New share issues ...	10	4	- 1	25	9	6
Retained earnings ..	9	9	4	19	8	- 7
Overall	7	7	3	17	8	- 5

10 per cent. domestic inflation and 0 per cent. world inflation

	Belgium	France	Germany	Italy	Nether-lands	United Kingdom
Source of finance						
Debt	-78	19	-	-	-93	1
New share issues ...	9	10	14	16	8	5
Retained earnings ..	18	14	8	32	14	- 3
Overall	21	14	16	43	17	- 2

- = undefined.

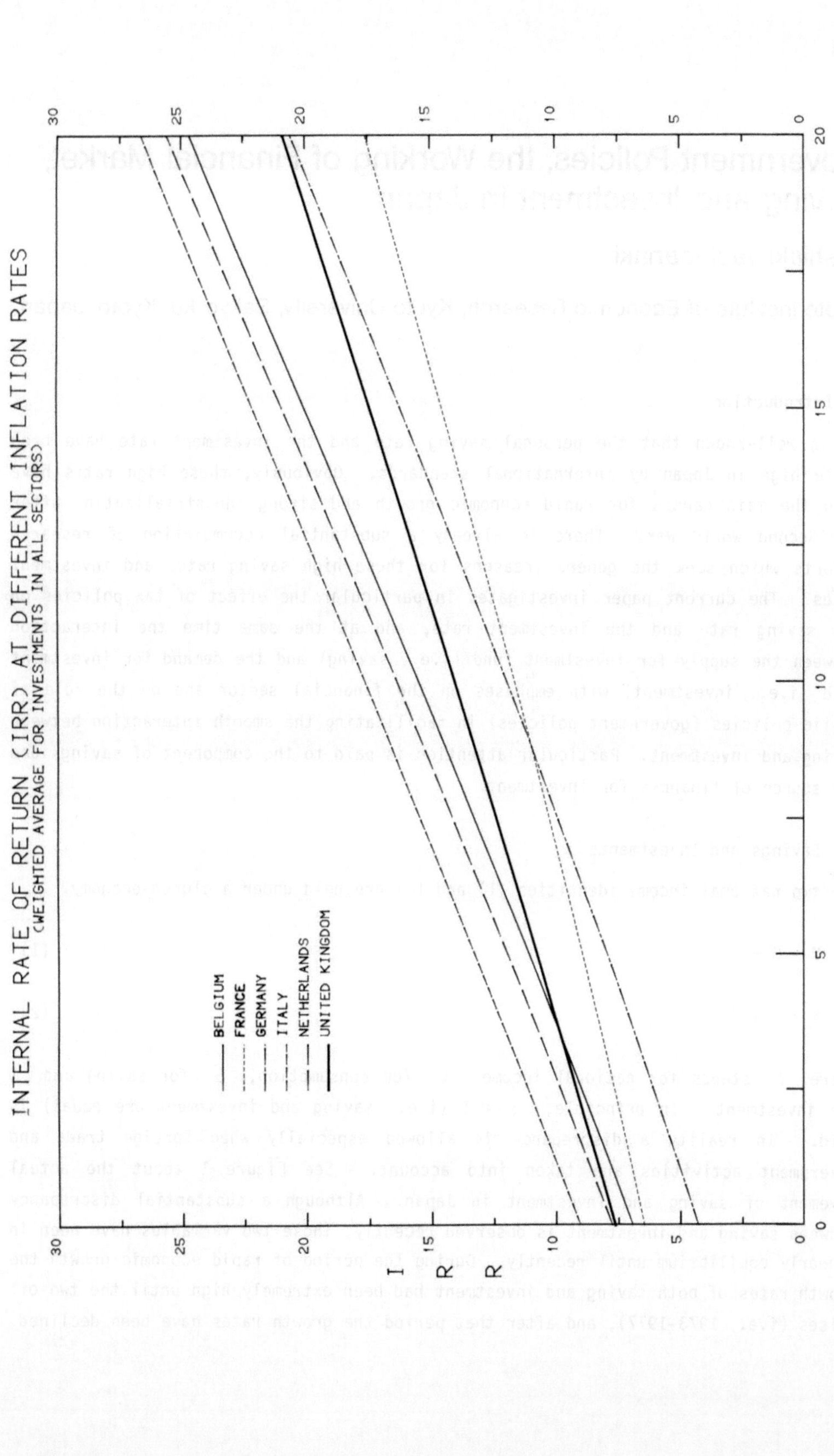
INTERNAL RATE OF RETURN (IRR) AT DIFFERENT INFLATION RATES
(WEIGHTED AVERAGE FOR INVESTMENTS IN ALL SECTORS)
BELGIUM
FRANCE
GERMANY
ITALY
NETHERLANDS
UNITED KINGDOM
IRR
INFLATION RATE
0
5
10
15
20
25
30

Government Policies, the Working of Financial Market, Saving and Investment in Japan

Toshiaki Tachibanaki

Kyoto Institute of Economic Research, Kyoto University, Sakyo-ku, Kyoto, Japan

I. Introduction

It is well-known that the personal saving rate and the investment rate have been quite high in Japan by international standards. Obviously, these high rates have been the main causes for rapid economic growth and strong industrialization after the Second World War. There is already a substantial accumulation of research efforts which seek the general reasons for these high saving rates and investment rates. The current paper investigates in particular the effect of tax policies on the saving rate and the investment rate, and at the same time the interaction between the supply for investment fund (i.e., saving) and the demand for investment fund (i.e., investment) with emphases on the financial sector and on the role of public policies (government policies) in facilitating the smooth interaction between saving and investment. Particular attention is paid to the component of savings and the source of finances for investment.

II. Savings and Investments

The two national income identities (1) and (2) are held under a closed-economy.

$$Y = C + S \tag{1}$$

$$Y = C + I \tag{2}$$

where Y stands for national income, C for consumption, S for saving and I for investment. In principle, S = I (i.e., saving and investment are equal) is held. In reality a discrepancy is allowed especially when foreign trade and government activities are taken into account. See Figure 1 about the actual movement of saving and investment in Japan. Although a substantial discrepancy between saving and investment is observed recently, those two variables have been in a nearly equilibrium until recently. During the period of rapid economic growth the growth rates of both saving and investment had been extremely high until the two oil crises (i.e., 1973-1977), and after that period the growth rates have been declined.

III. Components of Savings by Households and Sources of Corporate Finance

(i) International Comparison

Table 1 shows the composition of financial assets held by individuals for four countries, namely Japan, the U.S., the U.K. and FRG, while Table 2 shows the sources of corporate finance for the same countries. It is useful to look at Table 1 and Table 2 simultaneously because those two tables clearly present the difference in interactions (or channels) between the suppliers (i.e., ultimate lenders) of the fund and the demanders (i.e., ultimate borrowers) of the fund in the economies.

Several observations are possible based on Table 1 and Table 2 particularly from the view of international comparison. Most importantly, it is possible to classify the countries into two groups based on the structure of the financial market, namely (1) Japan and FRG, versus (2) the U.S. and the U.K. The reasons for proposing this classification are as follows. Japan and FRG have higher shares of currency and deposits than the U.S. and U.K., while the latters have higher shares of insurance & pensions, and securities in the composition of financial assets by individuals. In the source of corporate finances the shares of banks' loan are considerably high in Japan and FRG, while the roles of internal funds are more important in the U.S. and the U.K. than those in Japan and FRG. Incidentally, debentures in the U.S. and stocks (equities, or shares) in the U.K. have some significant roles.

Borrowing the traditional terminology in the monetary theory, for example from Gurley and Shaw (1960), the above classification may be characterized by the words, "indirect financing" versus "direct financing". Obviously, Japan and FRG belong to the former category because banks play an important role as financial intermediaries which collect funds from individuals (ultimate lenders) and finance corporate sectors (ultimate borrowers). Instead, the U.S. and U.K. corporations collect funds directly from ultimate lenders without having the indirect channels such as banks and others. In other words, ultimate borrowers sell their securities directly to ultimate lenders. Needless to say, neither Japan and FRG, nor the U.S. and the U.K. specialize completely in "indirect financing" or "direct financing". The only difference is the relative weight.

It is possible to examine the structure of the financial market both based on the ultimate lenders' side and based on the ultimate borrowers' side. We intend to examine the case of Japan about both lenders' and borrowers' behaviours to understand the working of the financial market in Japan which connects between savings and investments.

(ii) Portfolio Behaviours of Household Sector

First, the portfolio behaviours of ultimate lenders (in particular households) are examined. Table 3 presents a more detailed result of the composition of financial assets held by households than Table 1. The numbers in Table 3 are not strictly comparable with those in Table 1, partly because Table 3 contains savings at non-financial institutions, and partly because samples are somewhat different. Two important observations are possible. First, a very high share of demand deposits and time deposits is observed. Securities, including both stocks & shares and bonds & trust, show a lower rate. Secondly, when we pay attention to financial institutions with which individuals have transactions, it is recognized that postal offices have an important share as financial institutions. About 30-40 % of deposits and life insurance are handled by postal offices. These two facts are discussed in detail below in relation to the interaction between savings and investments.

A very high share of demand deposits and time deposits, and a low rate of securities which are led by household preferences determine the structure of the financial market to a greater extent. Since an extremely large share of funds owned by ultimate lenders is deposited at banks (both private banks and postal offices), the possibility of direct financing (i.e., ultimate lenders buy bonds and shares of corporate sectors and of government directly) is greatly reduced. There are no significant funds available for committing to buying bonds and shares since consumers have deposited their funds at banks based on their own initiatives. Banks and postal offices which have a large amount of funds lend them to corporate sectors, and also buy bonds and shares issued by corporate sectors and governments. This is the main mechanism of indirect financing in Japan. The role of postal offices will be discussed later.

The view mentioned above understands that the overwhelming presence of deposits at banks produced the Japanese way of indirect financing, and that corporate sectors were enforced to rely on it because only a small portion of funds is available for ultimate lenders to buy bonds and shares. Only a few people are willing to buy bonds and shares of corporations and governments directly. It is important to add two facts: first, the market for debentures (corporate bonds) was quite premature. In principle, borrowing from banks and issuing bonds are substitutable from corporate sectors' point view when the two transactions work and operate perfectly. In reality there have been significant restrictions to Japanese corporations in selling their debentures for various reasons. For example, guarantees (or collaterals) were required in exchange for issuing debentures. In the U.S. corporate bonds are, in principle, sold without collaterals. Secondly, the equity market also has been underdeveloped until quite recently. It is expanding rapidly nowadays, however. In sum, the only choice left for corporate sectors was to rely mainly on borrowing from banks willingly or unwillingly to finance their economic activities.

Before discussing the corporate sectors' view on corporate finance, we examine why households in Japan preferred deposits at banks and postal offices rather than the other financial goods when they faced a portfolio choice of financial assets. Since I concluded that the households' portfolio behaviours were responsible for producing the Japanese way of indirect financing, it is important to investigate why such unique behaviours of household sectors were observed.

Table 4 presents the before-tax rates of return to various financial commodities. It indicates that the average rates of return to shares (equities) and to bonds are considerably higher than those to time deposits. The rate of return to shares, when realized capital gains are included, is very high in many years. Of course, since it is a risky asset, the return is fluctuated considerably. As was given previously, people deposited much at banks and postal offices despite the low rates of return. Why do Japanese people prefer deposits to shares and bonds? First, one of the plausible answers is to raise the fact that Japanese households may have been quite risk-averse in the sense that they preferred <u>fixed</u> returns from savings even if their expected returns were lower. They wanted to avoid risky assets. It is noted, however, that some of higher income households seek high-return but high-risk financial assets recently. Secondly, banks and postal offices have an enormous number of branches all over the country which may be accessed by customers easily. It is likely that extremely hard competitions among banks raised the total amount of deposits more than light competitions would absorb. The services provided by banks were excellent. Thirdly, the knowledge and marketing possibility of bonds and shares was limited. In other words, households did not have sufficient informations on the bonds and share markets. Fourth, as will be explained later, the tax system is fairly favourable of deposits at banks and postal offices.

A risk sharing arrangement between banks and ultimate lenders must be mentioned in order to explain why ultimate lenders and banks had such good continuous customers relations. The indirect financing encourages this risk sharing device, as Ikeo (1985) points out correctly. According to Ikeo the main function of the indirect financing is the creation of "indirect securities". Ultimate lenders in principle have two possibilities for depositing their funds. One is to buy indirect securities sold by banks, and the other is to buy securities directly issued by ultimate borrowers. Since indirect securities are more liquid and less risky than direct securities, households (ultimate lenders) are inclined to buy indirect securities. Banks are willing to bear risks, and thus to hold direct securities issued by ultimate borrowers. In other words, households transferred their risks to banks by holding indirect securities.

This institutional arrangement of equity holding in Japan is quite different from those in the Anglo-American countries. While English and American banks are not allowed for holding shares of non-financial corporations in principle, Japanese banks are permitted to hold them up-to a certain limitation. In fact the greatest part of shares of non-financial corporations in Japan is held by financial institutions, in particular by private banks. See Aoki (1984) about this. Incidentally, the German case is similar to the Japanese one. This system obviously promotes the Japanese way of indirect financing, and supports banks as risk-bearing institutions. It is noted that private banks are not the only institutions which bear total risks. As Sasakibara, Feldman and Harada (1982) propose, the central bank (i.e., the Bank of Japan) and/or the Ministry of Finance shared risks as the final risk bearers. This will be explained later again.

(iii) Corporate Sectors' View on the Source of Finance

Next, the corporate sectors' (ultimate borrowers) view on the source of finance is reviewed. Although the overwhelming presence of funds deposited at banks was the primary cause for the indirect financing in Japan according to the present author's judgement, it is possible to argue that corporate sectors also have preferred the indirect financing.

Several merits of indirect financing from corporate sectors' view point are discussed here. The indirect financing in Japan is interpreted frequently by a word, the "main bank system". The main bank is a principal bank which has normally the highest rate of lending activity to a company and holds the largest equities of the company among all the banks, which lend funds to the same company and hold the equities of the same company. The main bank has a special and continuous relationship with the company. The merits of indirect financing can be explained easily by considering the merit of the main bank system.

First, the agency cost is saved considerably. The agency cost between lenders (banks) and borrowers (non-financial corporations) arises since there exists asymmetry of informations between them. It is costly for banks to examine the financial conditions and to monitor investment activities of firms. So is true for firms to provide all the financial conditions and investment plans with banks. By having the main bank system, those agency costs are saved considerably because the continuous customers' relationship does not require extra cost.

Furthermore, a main bank and a non-financial corporation continue their bilateral transactions under an implicit long-term contract and understanding. An implicit understanding assumes that "when the corporation is under a financial difficulty, the main bank gives several special helps to the firm in the following ways: (1) reduction in the rate of interest, (2) suspending the repayment of debt and (3) sending personnels as managers and others." Obviously, the firm has to respond to a favourable treatment given by the (main) bank by guaranteeing continuous transactions and paying sometimes implicit insurance preminum. This insurance scheme is not explicit but implicit. The bank implicitly understands that the insurance gives a special incentive for helping the firm when the customer is in a trouble. The above mechanism may be regarded as an implicit risk sharing device between a bank and a firm which has been quite common during the period of rapid economic growth. The mechanism has been eroded slightly in recent years.

There are several institutional arrangements or supplementary requirements to enhance this risk sharing device between banks and firms. For example, (1) compensating balances which require firms to deposit a part of funds, which were borrowed from the bank, at the same bank. These may be regarded as contributions to insurance which are paid by firms. (2) Concerted lending activities among the banks concerned, which decide the total amount of funds to be lended to the firm, and the share of each bank to be lended. This is another form of risk sharing especially for default-risks among private banks. (3) An important role of governmental financial institutions (frequently called Fiscal Investment Loan Programs) which bear defaul-risk. Private banks are willing to take risks since the financial institutions owned by government give an implicit credit to private banks. This will be argued again when we discuss about the postal saving. Related to this, the central bank (the Bank of Japan) also exists as a final risk bearer in the sense that the central bank has been always willing to help private banks. (4) Several long-term credit banks (which are not governmental banks) also created this risk sharing device and encouraged it. Those credit banks sell bonds, and lend funds to firms by using funds collected by their bonds. An important fact is that private banks bought the major part of bonds issued by the credit banks. This system diversifies the risk by creating the long-term credit banks as another source of financial intermediaries. See Teranishi (1982), Sasakibara, Feldman and Harada (1983), and Ikeo (1985) about the detailed discussions of the institutional arrangements of risk sharing devices in the Japanese banking system.

(iv) Summary

The Japanese way of indirect financing may be summarized by Figure 2. Ultimate lenders deposit their main part of funds at banks and postal offices (A), and buy bonds and shares very marginally through the security market (B). Financial sectors buy bonds and shares of ultimate borrowers through the security market (C), and also do lending activities to ultimate borrowers (D). This lending is very important. The combination of lines (A) and (D) may be called "credit-based system" according to Zysman (1983), or "loan-based system" according to Kaizuka (1977, 1983), while the combination of lines (B) and (C) may be called "capital market-based system" (Zysman), or "bond-based system" (Kaizuka). In the past especially during the period of rapid economic growth "loan-based system" was dominant, and "bond-based system" was relatively minor. This is one of the features of the indirect financing, and the so-called excessive borrowing from banks by corporate sectors was quite common. See, for example, Suzuki (1974). In recent years, however, "bond-based system" became important particularly after a large amount of the government bonds caused by the huge government budget deficit in the 1980s appeared in the financial market. More concretely, line (C) is important nowadays mainly because financial sectors buy bonds issued by public sectors (both central and local governments). It should be noted, however, that the share of corporate bond is still minor even recently. Line (B) also is showing an increasing trend gradually and gradually. Line (D) reduced their importance considerably. See Aoki (1984) who analyzed the changing power balance between individual investors and banks, which determines the whole structure of corporate finances and individual investments.

IV. Government Public Policies

This part examines whether the government policies were effective in promoting saving and investment in Japan. Four policy tools are available. One is to use financial institutions owned and managed by government fully. Second is to utilize tax policies explicitly. Third is to keep the rate of interest as low as possible to encourage investment. Fourth is to promote saving and investment by campaigning and giving administrative guidances to ordinary people and industries. This paper is concerned mainly with the first three instruments partly because the effect of the first three is evaluated quantitatively without much difficulty, while the fourth effect is not easily quantified. This does not necessarily imply, however, that the fourth effect was less important. It is ironical to note that the government has to campaign to promote consumption in recent years to reduce the I-S imbalance.

(i) Financial Institutions Managed by Public Authorities

Figure 2, which was explained previously, contains postal offices as one of the financial institutions. Postal offices are typical financial institutions owned by government. Since the postal bureau has an enormous number of branches, their ability of collecting deposits is extremely high. About 35-40 percent of the total time deposit held by households is deposited at the postal offices. Although this high share is often critisized by private banks as a business invasion of the public sector into the private economy, it is nonetheless an important source of the government revenue. These funds collected by the postal offices together with the funds accumulated by social security contributions are used as investment funds prepared by the government. In reality those investment funds (Zaisei-toushi) or Fiscal Investment Loan Program (FILP) are administered by semi-governmental financial institutions such as the Japan Development Bank, the Japan Export-Import Bank, the Banks for Medium/Small Firms, and others.

The Japan Development Bank played an important role in allocating these investment funds into several specific industries such as (1) electric power industry, (2) ship-building industry, and (3) coal mining industry which were regarded as the leading industries to restore the Japanese economy destructed by the War. The Japan Export-Import Bank also was crucial in allocating those investment funds into industries which promote export activities. See Ogura and Yoshino (1984) about the FILP. A controversy remains, namely the real effect of the FILP on investment.

Ogura and Yoshino (1983) calculated the reduction in interest payments for industrial sectors, when the FILP is used. Since the rate of interest financed by the FILP is lower than that financed by private banks, corporations which borrow funds from the FILP can save some cost of interest payment. Table 5 shows the estimated reductions. The largest reduction is given by marine transportation industries both for the period of rapid economic growth (1961-73), and for the post-oil crisis period (1974-80). The next important industries are transportation machinery (especially ship-building industries for the both periods), and electric power industries for the period of rapid economic growth. It is no exaggeration to describe that only the three industries, namely (1) marine transportation, (2) ship-building and (3) electric power industry, have benefitted much from the FILP, and that the other industries' benefit is very minor as shown by the figures for the total manufacturing. It should be noted, moreover, that those three industries are not the leading industries for strong industrialization but rather the declining industries (including coal mining industries and excluding electric power industries). In sum, quantitatively speaking, the contribution of the FILP to total cost saving is relatively minor.

It is possible, however, to conjecture that the public investment funds (i.e., the FILP) might encourage private investments for the other reason; the supply and the permission of those funds to a particular industry (or a firm) by the FILP may give a positive signal to that industry or firm because it may be regarded as receiving a guarantee from the government and as a sound industry or a firm. Sakakibara, Feldman and Harada (1982), Hyugano (1984) and others stress this effect. Several empirical studies such as Ogura and Yoshino (1985), and Horiuchi and Ohtaki (1987) do not accept this effect.

In sum, it is not still convincing to believe that the public investment funds were effective in encouraging private investments and in promoting strong industrialization. A controversy has not been settled yet with respect to the impact of governmental financial institutions on private investment.

(ii) Low Interest Policies

It is often said that Japanese policy authorities adopted, "a low interest rate policy," and this policy enabled the Japanese economy to perform relatively well. The reason is that it provided the private sector with higher investment incentives. See Teranishi (1982), or Kosai and Ogino (1984).

It is important to define which interest rate has been kept lower artificially by the public authority among different interest rates; (1) official discount rate by the Bank of Japan, (2) interest rate on bank deposits, (3) money market rate (inter-bank short-run loan rate) called "call money rate" in Japan, and (4) bank lending rate (or bank loan rate)? Also, it is important to distinguish, whether it is the nominal or the real rate, or whether it is the effective rate or not. Horiuch (1984) performed a useful international comparison with respect to both the nominal and real interest rates for the three interest rates, namely (1) discount rate, (2) bank deposit rate, and (3) call money rate. He found that the real interest rates (both discount rate and call money rate) in fact had been higher in Japan than those in the U.S., the U.K. and FRG, although only the bank deposit rates were kept lower by the administrative regulations.

There is another important observation related to the interest rate; the regulation adopted by public authorities not only lowered the level of bank loan rates, but also kept its fluctuation at a minimum level. See Figure 3. It is believed that the bank loan rate which was kept lower artificially facilitated relatively cheaper funds for investment to industrial sectors. Horiuchi (1984) gives us a doubt about this interpretation, by pointing out following reasons. First, as many authors suggest, the effective bank loan rate was significantly higher than the nominal rate because compensating balances which require firms to deposit a part of the loan at the bank were considerably common in the Japanese banking activities. Secondly, the relatively low levels of the discount rate and deposit rates together with the less flexible bank loan rate were not effective in transfering a substantial amount of subsidies to industries and in giving benefits to ultimate borrowers (industries) in the form of low interest rates. Horiuchi speculates, by presenting the fact that after-tax profit rates were higher in banking sectors than those in non-banking sectors, that private banks themselves benefitted highly. It is the private banks who gained the largest benefit under the regulated interest rates. I share with the above opinion, and would like to add another fact: the banking industries and the other financial institutions were protected strongly by public authorities (both the Ministry of Finance and the Bank of Japan). In order to avoid bankrupcies of financial institutions which hurt consumers who deposit their funds at those financial institutions, the central government was always willing to help private financial institutions and to provide them with considerable credits. Private banks were able to give excessive bank loans to firms when there were demands, because they are aware of the fact that the central authority would bear the final risk. The above conjecture is supported by recognizing that there have been no virtual bankrupcies of private financial institutions in Japan after the World War II. Of course, there have been a considerable number of business troubles and near bankrupcies of smaller size private financial institutions. On those occasions the Ministry of Finance and/or the Bank of Japan took a positive action in providing special credits or suggesting mergers with the other financial institutions in order to avoid bankrupcies.

A lower bank loan rate may be interpreted as an example of an adverse selection due to the asymmetric informations between lenders (banks) and borrowers (firms) à la Stiglitz and Weiss (1981). Ikeo (1985) pointed out that banks would have been able to keep the bank loan rate lower even without the government intervention for following reasons. It would be preferable for banks to have a higher bank loan rate, if the default risk was very small. It is, however, possible to conjecture that a firm, which is willing to accept a higher bank loan rate, may have a higher default risk. Since a bank does not have perfect informations on the firm's current condition, it may be desirable to keep the bank loan rate lower or to consider a credit rationing in order to attract borrowers whose default risk is small. It is possible that a higher bank loan rate would discourage the borrowing activity of a firm which has a lower default risk (i.e., the adverse selection). If this interpretation was accepted, it would be possible to conclude that the lower bank loan rate was preferred also by private banks. The government low interest policy was Pareto-improving, and banks accepted this policy willingly.

Ikeo (1985) gave a new interpretation to the notorious "compensating balance," which is often criticized as an unfair transaction practice. Under the assumption of asymmetric informations a bank can use the level of compensating balance accepted by a borrower as a signal of the borrower's default risk. La raison d'être of the compensating balance was proposed, and compensating balances are not necessarily unjustified practices.

It would be a future task to determine whether the government policy (for example, a low interest policy) or the traditional banking practice (for example, compensating balances) was consistent with the natural market-enforcement without the government intervention, or was desirable for both banks and borrowers (firms).

(iii) Tax Policies

It is an attractive subject to examine whether tax policies are effective in encouraging personal savings and corporate investments in view of the fact that the Japanese tax systems apparently aimed at achieving such goals. In fact there is a considerable number of research efforts in Japan in this area which examine the real effects of tax policies. This section attempts to give a brief survey and to point out several unsolved issues. The effect on saving and investment is discussed separately, although some issues are interrelated.

(a) Saving

It is no exaggeration to conclude that the Japanese personal tax system of interest incomes and dividend incomes was quite generous in hoping that the system encourages more saving since higher savings could be used for higher rate of investments. See, for example, Shoven and Tachibanaki (1986) about the description of the Japanese personal income tax system which is aiming at encouraging saving.

The Japanese economic development since the War confirms a story such that higher savings produced higher investments. However, a story such that the generous personal income tax system of interest and dividend incomes has encouraged more saving is not supported empirically. The classical article by Komiya (1966) proposed that there was no strong evidence to support that the Japanese saving rate was raised by the generous tax system. Since this classical investigation appeared, no investigations have changed the understanding of the no effect of tax polices on personal savings, although the number of the investigations is not large.

It is necessary to discuss about why the effect of tax policies (especially, low tax rates on interest and dividend incomes) was negligible in raising the saving rate. First, the saving elasticity with respect to the rate of interest is very small, as pointed out previously. A large number of studies of time-series consumption or saving functions except for perhaps Turner (1986), and Tachibanaki and Suruga (1987) report that the effect of the interest rate on savings is statistically insignificant. If one believes that the saving elasticity with respect to the interest rate is very small, it is quite natural that the rate of saving is unaffected by a tax on interest (and dividend) incomes. Horioka (1987) supports it even in the cross-section regional data. Secondly, somewhat related to the above, the effective tax rate on returns from financial assets also is very small. This is true for all income classes. See Table 6 about the figures of the effective tax rates by income classes. Table 6 shows that there is no significant tax progressivity on returns from financial assets. If this was true, it would be possible to conceive of the fact that the saving rates (especially the saving rates for financial assets) would be higher in higher income classes than those in lower income classes. However, as Table 6 indicates, the difference in the saving rates for financial assets by income classes is not so great as it would be anticipated, although the difference in the total saving rates by income classes is somewhat larger. This is an interesting myth.

Ishikawa (1987) performed an interesting simulation, proposing that the saving rate would not be increased significantly even if the current favourable tax treatment of interest and dividend incomes was abolished. At the same time, he presented that although an increase in income transfers from higher income classes to lower income classes caused by such an abolishment was observed under the assumption of revenue neutrality, it did not necessarily increase the average rate of personal saving.

Summarizing the above discussions, it is hard to believe that the favourable tax treatment of financial assets (interest and dividend incomes) was effective in raising the personal saving rate. A potentially important aspect is the low level of the average tax (especially income tax) burden and of social security contributions. The average Japanese tax and social security burdens have been considerably lower than the other industrial nations. A low level of the average income tax rate gives households more room for higher saving, since the disposable income is not greatly reduced. A low level of social security contributions encourages higher saving, if the social security system is regarded as a substitute for private saving. See Sasaki and Tachibanaki (1985) who showed a small but non-negligible possibility of the substitution effect in Japan.

Another important aspect of the effect of tax policies on savings is the effect on the composition of financial assets rather than on the aggregate saving rate. Fujita (1972) pointed out a possible effect of the change in tax policies on the change in the composition of financial assets. It is not so easy to investigate this issue thoughtfully because the taxation on various financial assets is extremely complicated. For example, some assets are completely free from any taxations, while some are taxed separately as withholding taxes and some are taxed within a framework of comprehensive income taxations. Capital gain taxes and local taxes make the taxation of financial assets more complicated. In other words, it is hard to estimate the effective tax rate on each financial asset. This difficulty may be the main cause for the lack of empirical studies with respect to the effect of tax policies on the composition of financial assets.

With respect to this, Yoshino (1984) is a worthwhile attempt which investigated the effect of the tax free saving account on the composition of financial assets by five income classes. The Japanese tax law allows a certain amount of tax-free time deposits for each person. He proposed that households of lower income classes are sensitive to the change in the amount of the tax free saving account when they select various financial assets as forms of savings. Although the Yoshino's work is useful, it is not possible to draw a concrete conclusion since the study is based only on the grouped five income classes. An analysis based on individual observations is called for in order to investigate fully the effect of interest rates and of tax policies on the composition of financial assets.

As of April 1988 the tax system on returns to financial assets was considerably modified. Some changes in the composition of financial assets are anticipated due to this tax reform. We have to wait data availability to draw a concrete conclusion about it.

One potential problem in the analysis of the effect on the composition of assets is the ignorance of the effect of real assets. Households normally face two opportunities in allocating their funds. One is to allocate them between financial assets and real assets, and the other is to allocate the aggregate financial assets into various financial assets. It is somewhat risky to draw a concrete conclusion about the effect of tax policies on the composition of financial assets without investigating the relationship between real assets and financial assets, in view of the current Japanese economic conditions such as increasing land prices and a peculiar tax system on land.

(b) Investment

This section gives an overall review of the effect of tax policies on investment behaviours in Japan. It is no exaggeration to describe that the extremely aggressive investment activity was one of the causes for rapid industrialization after the War, although its investment activity has been eroded considerably after the oil-crises. It is an appealing subject to inquire whether tax policy was responsible for encouraging more investment. There are two separate issues when the effect of tax policies on investment activities is examined. First is to investigate the effect of tax policies on cost of capital. Second is to examine whether changing the cost of capital affects on the amount of investment. Although several attempts have been made for the first subject, the second subject has not been investigated fully yet.

The first subject has two approaches. One is concerned only with the effect of the taxation on cost of capital of corporate sectors, while the other is concerned with the interaction between personal income tax and corporate tax, and thus with the effect of the interaction on cost of capital. The first approach was adopted by Tajika and Yui (1984), Noguchi (1985), Takenaka and others (1986), Tajika, Hayashi and Yui (1987), and others, while the second approach was adopted by Shoven and Tachibanaki (1986), Kikutani and Tachibanaki (1986), Iwata, Suzuki and Yoshida (1987), and others.

The first approach is examined initially. The Japanese corporate tax law prepared several special measures such as (1) tax free reserves like (a) reserves for bad debts, (b) reserves for retirement allowance, (c) reserves for bonus payments, (d) reserves for price fluctuation, and several others, (2) tax allowance for inventory and depreciation, (3) investment tax credit, (4) cost reduction due to paying enterprise tax (one of local taxes), and others. See, for example, Shoven and Tachibanaki (1986) about these measures. Although some conflicting results were obtained among the above studies, several general observations may be summarized as follows.

(1) The reduction in corporate income tax due to tax free reserves was 7.1 percent on average between 1963 and 1980. Variations by industries about those effects are considerable. The highest reduction is public enterprises, and banking and insurances. Agriculture, iron and steel, printing and chemical industries follow. The reduction in cost of capital explained by these tax free reserves is only about 2-4 percent in total industries on average between 1963 and 1982, and 1.5-2.5 percent in manufacturing. Although some minor variations by industries are observed, the average rate of reduction is rather small. See Tajika and Yui (1984), and Ogura and Yoshino (1987) about these.

(2) Cost of capital which is explained by tax free reserves, corporate tax, the relative price between capital goods and outputs, depreciation, and real interest rate (nominal interest rate minus inflation rate), was estimated by Tajika, Hayashi and Yui (1987). It was found that the costs of capital in mining, construction, retail trade and service industries had been higher than in manufacturing industries in most years. Manufacturing industries, however, are higher than those industries except for construction industries in recent years. Fluctuations in the costs of capital by years are considerable, and they are attributable mainly to the change in real interest rates (and thus inflation rate also) and the relative price between capital goods and output.

(3) Takenaka and others (1986) estimated the costs of capital by adopting the idea of rental price of capital or user cost of capital, and attempted a decomposition analysis in order to investigate what factor among several factors such as (1) relative price, (2) discount rate, and (3) tax system, was important for explaining fluctuations in the rental price of capital. See Table 7. They obtained the similar results as Tajika, Hayashi and Yui (1987), namely, the most important factor was the change in relative prices, and the tax factor was very minor.

(4) Takenaka and others (1986) performed a valuable international comparison. Table 7 suggests that Japan shows a decreasing trend of rental price of capital while the U.K., the U.S. and FRG show increasing trends in general. The Japanese decrease is attributed mainly to the decrease in relative price, while the British and German increases are attributed mainly to discount factors. However, a tax factor in the U.K. fluctuates a lot. In the U.S. the main reason for an increasing trend is a discount factor (i.e., largely interest rate). At the same time, a tax factor and a relative price factor contributed to considerable fluctuations in the rental price of capital.

Summarizing the above examinations, it is concluded that the real effect of various tax policies aiming at reducing the burden of corporate sectors in Japan was small when the effect was estimated in the form of either cost of capital or rental price of capital. This is confirmed not only by the industrial studies in Japan but also by the international comparison.

The story, however, is considerably different when the interaction between personal income tax and corporate tax is taken into account in estimating the effect of tax policies. This is the subject of the second approach. Shoven and Tachibanaki (1986), Kikutani and Tachibanaki (1986), Iwata, Suzuki and Yoshida (1987) are representative studies which estimated the effective tax rates of capital income, based mainly on the King and Fullerton international study (1984). The principal intention of this approach is to consider the tax "wedge" between the rate of return on investment and the rate of return on saving, and to estimate this tax "wedge." Since the calculation includes the <u>post-tax</u> rate of return to the <u>saver</u>, a tax treatment of suppliers for saving plays an important role as well as a tax treatment of corporate sectors. In this calculation the low tax rate on savings on average is crucial for estimating the effective tax rate of capital income, although it was concluded previously that the low tax burden on savings had not necessarily encouraged higher aggregate savings.

It is concluded based on Kikutani and Tachibanaki (1986) that the estimated effective tax rates on income from capital in 1961-1980 in Japan were considerably lower than those in the other countries such as the U.K., Sweden, FRG and the U.S. for following two reasons. (1) A higher reliance on debt-financing in Japan lowers the effective tax rate of capital income based mainly on the corporate sector's contribution. The crucial element is tax deductibility of interest paid. (2) Much lower effective tax rates on interest and dividend incomes contribute to lowering the estimated effective tax rates of capital income. Iwata, Suzuki and Yoshida (1987), however, presented somewhat different results in 1983 between Japan and the U.S., namely, the U.S. effective tax rate is slightly higher than Japan. There are two reasons for this difference between the two studies. First is the difference in the observation years. The U.S. had a major tax reform in 1981. Second is that the treatment of economic depreciation rates is different between the two. Nevertheless, the two studies propose the same conclusion; the lower effective tax rates on interest and dividend incomes for savers in Japan are responsible for lowering the effective tax rates of income from capital in Japan. This is an important observation, and gives us a policy implication such that tax policies for savers may be used as useful weapons for reducing the burden of capital income and possibly for encouraging more investment.

Now, we come to the final subject; the effect of the change in cost of capital on investment behaviours. Two representative studies in this field, namely Homma and others (1984), and Ogura and Yoshino (1987), gave somewhat conflicting results about the evaluation of the effect of tax policies on investment. Homma and others (1984) estimated the tax-adjusted q-type investment functions by industries for time-series data, and obtained a policy implication such that a reduction in tax burders of corporate sectors would increase the amount of investment substantially. At the same time, it was found that the effect of an investment tax credit was far more significant in raising investment than that of a reduction in the corporate tax rate. Ogura and Yoshino (1987) estimated a modified version of the Uzawa (1969) and Stiglitz (1973) types investment functions. Their investment function which is explained by excess profit (i.e., revenue minus cost, and both are functions of several tax parameters) and several lagged investments was fairly successful. It was proposed that it would be impossible to conclude that both the reduction in tax burdens of corporations due to inflation, and several special measures of tax policies such as special depreciation allowances aiming at increasing investment activities were effective in raising the amount of investment.

Since the above two studies adopt considerably different techniques in deriving the quantitative effect of tax policies on investment, the conflicting results between the two are not surprising. Future works are called for. In particular, it is necessary to investigate the effect of cost of capital, which takes account of the interaction between personal income tax and corporate tax, on investment because all the studies of the effect of tax policies dealt only with corporate tax, and ignored the contribution of personal income tax. A new and different finding might be obtained when the contribution of personal income tax to the influence of the cost of capital on investment activities was taken into account.

V. Concluding Remarks

The paper investigated two issues; the first one was to investigate the interaction between savings and investments in Japan with a particular emphasis on the financial market which connects between ultimate lenders and ultimate borrowers. The Japanese way of financing was compared with the other industrial countries both for household behaviours (i.e., ultimate lenders) and corporate behaviours (i.e., ultimate borrowers), and its characteristics were discussed.

It should be emphasized that the financial market in Japan is changing drastically in recent years, as represented by (1) internationalization, (2) securitalization, (3) de-regulation, (4) globalization and some others. We need another piece of work, which takes account of those changes, in order to discuss and evaluate the effect of those changes on the working of the financial market.

The second issue was to examine the effect of various public policies, including the role of government financial institutions (the Fiscal Investment Loan Program), a low interest policy, and tax policies, on investment activities.

Table 1, Compositions of Financial Assets owned by Households in 1984 (%)

	Japan	U.S. (1983)	U.K.	FRG
Currency & Sight Deposit	11.0	5.8	36.6 (Currency & Sight Deposit and Time Deposit combined)	7.6
Time Deposit	59.7	31.7		47.7
Insurance & Pension	15.7	25.5	39.5	19.5
Securities (Stocks & Bonds)	13.1	35.1	17.1	17.5
Stocks	(1.5)	(22.6)	(12.5)	(1.8)
Total	100.0	100.0	100.0	100.0

Notes

(1) The definitions of financial assets and the classifications are somewhat different by countries.

(2) Source: International Statistics published by the Bank of Japan.

Table 2, Source of Corporate Finances in 1979-83 (average figures) (%)

	Japan	U.S.	U.K.	FRG
Internal Fund	48.3	62.7	66.6	43.6
Stocks (Shares)	3.2	1.9	4.1	1.4
Debentures	2.5	8.9	1.2	0.1
Borrowing from outside				
Private Banks	27.5	12.1	17.5	24.9
Public Banks	3.9	0.3	0.2	5.6
Inter-firm Credits	14.5	9.1	-	1.5
Others	0.1	5.0	32.0	23.0
Total	100.0	100.0	100.0	100.0

Notes
(1) Internal fund normally means the sum of depreciations and retained earnings.
(2) Public Banks in FRG include various kinds of financial institutions such as pension funds, life insurance companies, banks for housing construction and others.
(3) Source: Flow of Funds Tables prepared by the central banks.

Table 3, Composition of Financial Assets held by Households in 1985 (%)

Saving	100.00		
(Financial institutions)	96.92		
Demand deposits		7.61	
Post offices			1.38
Banks			4.02
Others			2.22
Time deposits		47.08	
Post offices			14.27
Banks			19.86
Others			12.96
Life insurance		22.61	
Postal life insurance & Postal annuity			8.27
Life insurance company			14.34
Securities		19.61	
Stocks & shares			8.30
Public & corporate bonds			4.48
Unit & open-end trust			0.93
Open-end bond trust			1.22
Loan trust & money in trust			4.68
(Non-financial institutions)	3.08		
Deposits in one's own companies		2.16	
Mutual financing associations		0.05	
Others		0.87	

Source: Family Savings Survey 1985, published by the Prime Minister's office

Table 4, Before-tax Rates of Return to Financial Assets (%) per year

Financial Assets	1980	81	82	83	84
Deposits at Post office					
Fixed Amount Deposits	5.62	6.67	6.65	6.93	6.64
Demand Deposits	4.25	3.74	3.14	3.12	2.88
Deposits at Banks					
Time Deposits	7.15	6.30	5.54	5.51	5.25
Demand Deposits	2.93	2.39	1.78	1.75	1.51
Shares					
Including Capital Gains	8.50	21.00	2.10	23.00	29.40
Only Dividends	7.89	9.11	8.64	9.25	8.85
Deposits at Own Companies	6.94	7.26	7.49	7.44	7.44
Bonds	8.55	7.99	7.76	7.48	6.99
Trusts	8.17	7.52	7.16	7.16	6.18

Source: Various Documents of Financial Statistics

Table 5, Reduction in Interest Payments by Utilizing Public Investment Funds and Increase in Profit by Applying Special Depreciation Allowance (%)

	1961-73			1974-80		
	(A)	(B)	Total	(A)	(B)	Total
Manufacturing	1.0(52)	0.95(48)	1.95	1.6(73)	0.57(27)	2.1
Transportation-Machinary	6.9(85)	1.3(15)	8.2	6.4(94)	0.4(6)	6.8
Iron and Steel	0.5(27)	1.4(73)	1.9	1.6(70)	0.7(30)	2.3
General Machine	3.6(72)	1.4(28)	4.9	2.6(84)	0.5(16)	3.1
Marine Transportation	22.3(76)	7.2(24)	29.5	17.0(81)	4.1(1.9)	21.1
Electric Power	5.9(94)	0.36(6)	6.26	2.8(77)	0.9(24)	3.7

Notes

(1) (A) signifies the proportion of the reduction in interest payments to total fixed investment (including land), and (B) signifies the proportion of the profit due to special depreciation allowance to total fixed investment (including land), respectively.

(2) The number in parentheses is the share to the total.

(3) Source: Ogura and Yoshino (1983).

Table 6, Rates of Return to Various Financial Assets and Effective Tax Rates on Financial Assets by Income Class

Income Class	Gross Saving Rate (%)	Financial Saving Rate (%)	Before-Tax Rate of Return to Financial Assets (%)	After-Tax Rate of Return to Financial Assets (%)	Effective Tax Rate on Return (%)
First quintile	9.1 (34.3)	4.5 (38.6)	3.24 (1.84)	3.24 (1.84)	0.00 (0.00)
Second	13.4 (27.0)	6.1 (39.1)	3.48 (1.77)	3.47 (1.77)	0.04 (0.54)
Third	16.3 (26.5)	5.5 (44.7)	4.01 (1.92)	3.99 (1.92)	0.58 (3.03)
Fourth	19.6 (26.6)	9.2 (40.4)	4.72 (2.07)	4.50 (2.06)	5.15 (6.03)
Top 5 percent	21.2 (29.3)	8.9 (41.5)	5.18 (2.12)	4.80 (2.14)	8.79 (6.03)
Total	14.7 (28.6)	6.3 (40.8)	3.88 (1.91)	3.81 (1.90)	1.52 (3.48)

Notes
1) Numbers in parentheses are standard deviations
2) Source: Ishikawa (1987)

Table 7, A Change in Rental Price of Capital And Its Decomposition into Three Factors, (1) Relative Price between Capital Goods and Output, (2) Discount Factor and (3) Tax Factor (% change per year)

	Period	Relative Price	Discount Factor	Tax Factor	Rental Price
JAPAN	1965-1970	-4.51	3.67	-0.10	-0.94
	1970-1975	-1.41	-1.29	0.69	-2.01
	1975-1980	-1.27	-0.97	0.02	-2.22
	1980-1984	-2.27	-1.46	-0.24	-3.97
	1965-1984	-1.67	-0.34	-0.12	-1.89
U.S.A.	1965-1970	-0.13	3.21	3.88	6.96
	1970-1975	0.98	0.94	-1.09	0.83
	1975-1980	-0.14	3.25	0.24	3.35
	1980-1983	-2.65	0.81	-2.58	-4.42
	1965-1983	-0.36	2.18	0.02	1.84
FRG	1966-1970	0.51	1.51	0.57	2.59
	1970-1975	-1.00	2.93	-0.54	1.39
	1975-1980	0.00	2.03	0.56	2.59
	1980-1981	0.16	-2.34	-1.43	-3.61
	1966-1981	-0.54	2.34	0.10	1.90
U.K.	1966-1970	-0.84	1.94	0.24	1.34
	1970-1975	2.28	-0.21	2.54	4.62
	1975-1980	-0.60	1.92	0.39	1.71
	1980-1982	-3.89	-0.28	0.53	-3.64
	1966-1982	-0.28	1.10	1.09	1.91

Notes

(1) Source: Takenaka and others (1986).

(2) Calculations are made as follows:

$$C = \frac{p_k}{p} \{(1-\tau)i + \delta\} \frac{1-k-z}{1-\tau} = \mathbb{P} \cdot \mathbb{D} \cdot T$$

$$\frac{dC}{C} = \frac{d\mathbb{P}}{\mathbb{P}} + \frac{d\mathbb{D}}{\mathbb{D}} + \frac{dT}{T}$$

$\mathbb{P}$: relative price between capital goods and output

$\mathbb{D}$: discount factor

T: tax factor

C: rental price of capital

(3) Capital gains are not considered.

Figure 1, Movements of Saving Rates and Investment Rates in Japan

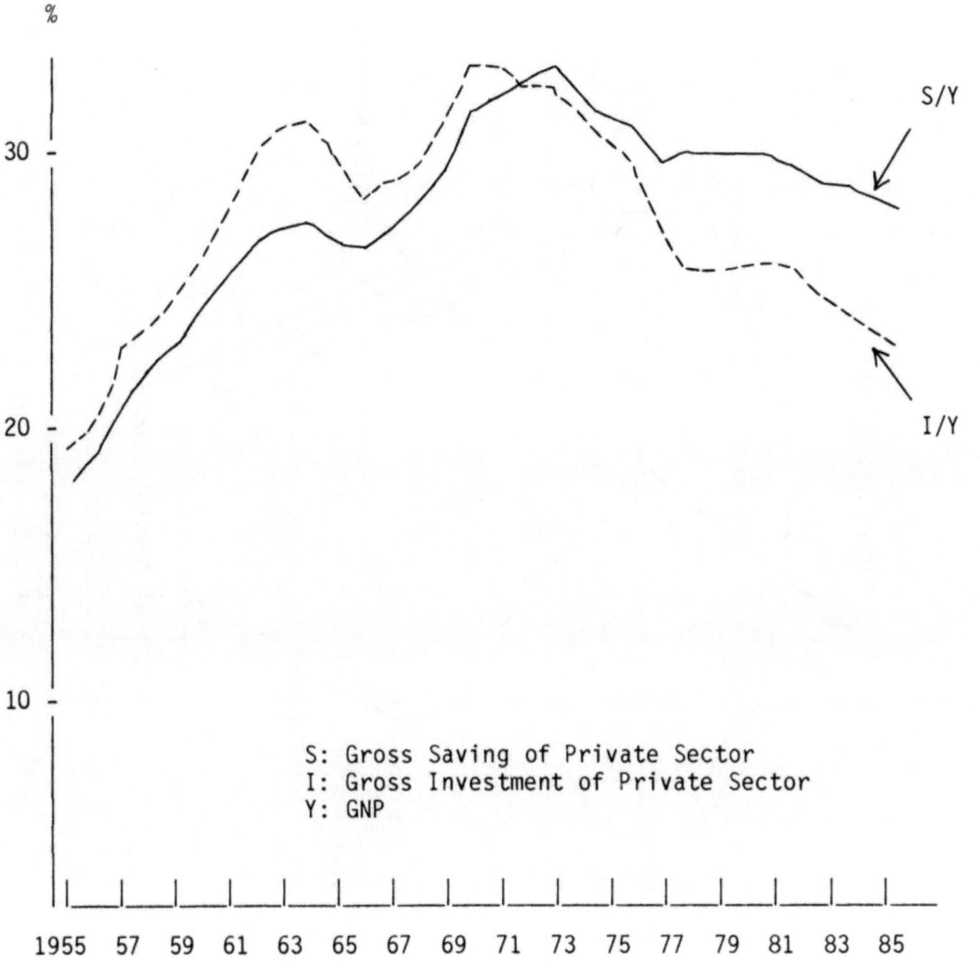

Source: National Income Statistics.

Figure 2, Japanese Way of Indirect Financing

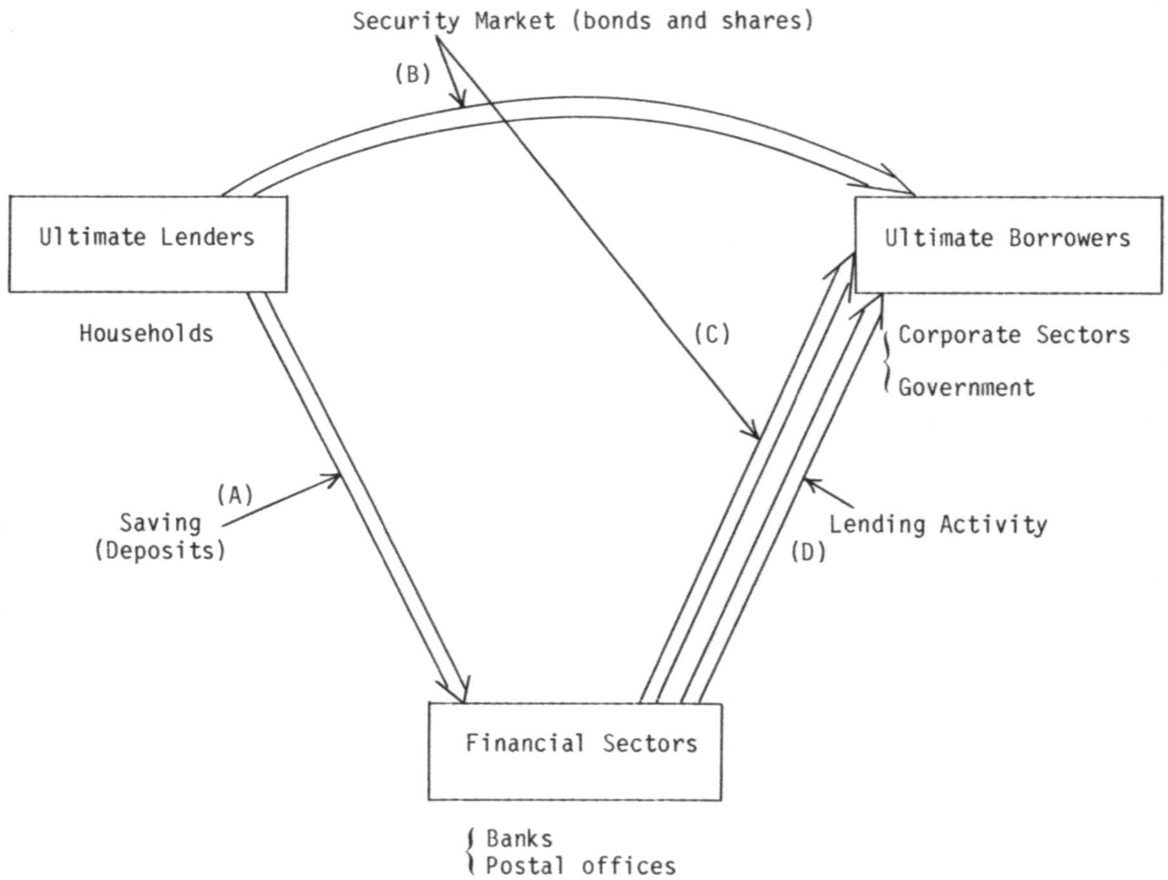

Figure 3, Movements of Various Nominal Interest Rates in Japan (1951-1980)

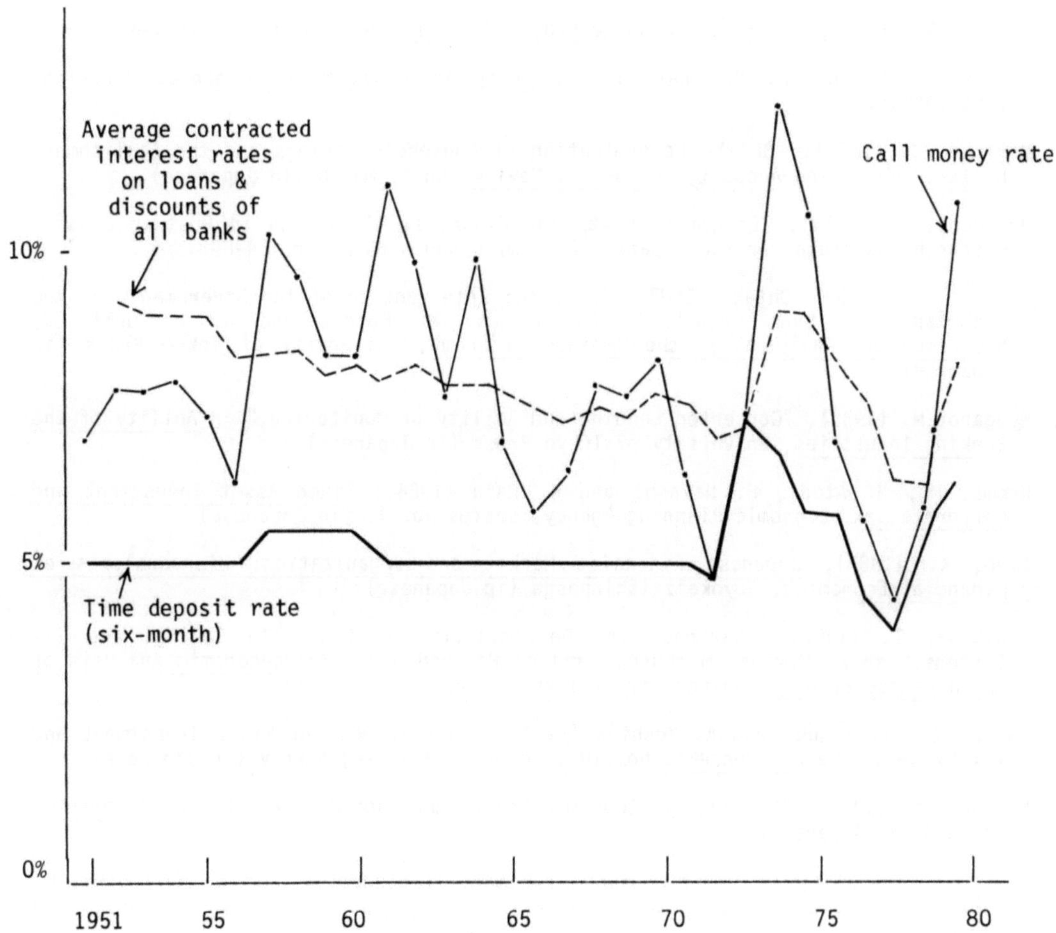

Source: Horiuchi (1984)

References

Aoki, M. (1984), "Shareholders' Non-Unanimity on Investment Finance: Banks vs. Individual Investors," in M. Aoki ed., The Economic Analysis of the Japanese Firm, North-Holland.

Fujita, S. (1972), Theory of Japanese Public Finance, Keisoshobo (in Japanese).

Gurley, J. G. and E. S. Shaw (1960), Money in a Theory of Finance, Brookings Institution.

Horioka, C. (1987), "On the Determination of Household Savings and the Abolishment of Tax Free Saving Account," Financial Review, No.4, March (in Japanese).

Horiuchi, A. (1984), "Economic Growth and Financial Allocation in Postwar Japan," Research Institute for the Japanese Economy Discussion Paper, 84-F-3.

__________ and M. Ohtaki (1987), "Finance; Intervention of the Government and the Importance of Bank Loan," in K. Hamada, M. Kuroda and A. Horiuchi ed, Macroeconomic Analysis of the Japanese Economy, University of Tokyo Press (in Japanese).

Hyugano, M. (1986), "Concerted Lending and Ability of Monitoring," in Ability of the Banking Industries, University of Tokyo Press (in Japanese).

Homma, M., N. Atoda, F. Hayashi and K. Hata (1984), Fixed Asset Investment and Corporate Tax, Economic Planning Agency, series No. 41 (in Japanese)

Ikeo, K. (1985), Japanese Financial Market and Organization: Microanalysis of Financial Economics, Toyokeizaishinposha (in Japanese).

Ishikawa, T. (1987), "Savings: The Determination of Household Savings and Tax Systems," in K. Hamada, M. Kuroda and A. Horiuchi ed., Macroeconomic Analysis of the Japanese Economy, University of Tokyo Press (in Japanese).

Iwata, K., I. Suzuki and A. Yoshida (1987), "Capital Cost of Fixed Investment and Tax Systems," Keizai Bunseki, No. 107, Economic Planning Agency (in Japanese).

Kaizuka, K. (1977), "A Change in Economic Environment and Monetary Policy," Jeristo, No. 628 (in Japanese).

__________ (1983), "Classifications of Financial Economies," in M. Moriguchi, M. Aoki and T. Sawa ed., Structural Analysis of the Japanese Economy, Sobunsha (in Japanese).

Kikutani, T. and T. Tachibanaki (1986), "The Taxation of Income from Capital in Japan: Historical Perspectives and Extentions," forthcoming in C. Bosworth and C. Hulten ed., The U.S. and Japan Productivity Difference, University of Chicago Press.

Komiya, R. (1966), "Post-war Japanese Tax System and Capital Accumulation," Todai Keizai Ronshu, Vol. 32, No. 2, July, pp. 10-35 (in Japanese).

Kosai, Y. and Y. Ogino (1984), The Contemporary Japanese Economy, Macmillan Press.

Noguchi, Y. (1985), Planning of Tax Reforms, Toyokeizaishimposha (in Japanese).

Ogura, S. and N. Yoshino (1984), "Tax System and the Fiscal Investment Loan Programs," in R. Komiya, M. Okuno and K. Suzumura ed., Industrial Policy in Japan, University of Tokyo Press (in Japanese).

__________ (1985), "Special Depreciation, Fiscal Investment Loan Program and Industry Structure in Japan," Keizaikenkyu, Vol. 36, No. 2, pp. 110-120. (in Japanese).

__________ (1987), "Tax System: Corporate Tax and Firm Behaviour," in K. Hamada, M. Kuroda and A. Horiuchi ed., Macroeconomic Analysis of the Japanese Economy, University of Tokyo Press (in Japanese).

Sasaki, M. and T. Tachibanaki (1985), "The Effect of Public Pension on Personal Saving Rates and Asset Accumulation by Cohorts," Shakaihosho Kenkyu, vol 21, June (in Japanese).

Sakakibara, E., R. Feldman and Y. Harada (1982), The Japanese Financial System in Contemporary Perspective, U.S. Government Printing Office.

Shoven, J. and T. Tachibanaki (1986), "The Taxation of Income from Capital in Japan," forthcoming in M. Moriguchi and J. B. Shoven ed., Government Policy Towards Industry in the United States and Japan, Cambridge University Press.

Stiglitz, J. (1973), "Taxation, Corporate Financial Policy and the Cost of Capital," Journal of Public Economics, Vol. 2, pp. 1-34.

__________ and A. Weiss (1981), "Credit Rationing in Markets with Imperfect Information," American Economic Review, Vol. 71, pp. 339-45.

Suzuki, Y. (1974), Contemporary Japanese Monetary Economy, Toyokeizaishimposha (in Japanese).

Tachibanaki, T. and T. Suruga (1987), "Consumption Function and Interest Rate," mimeo.

Tajika, E. and Y. Yui (1984), "Post-war Corporate Taxation and Fixed Investment in Japan: Estimation of Reduction in Corporate Tax by Industries," Gendai Keizai, No. 59, Autumn (in Japanese).

__________, F. Hayashi and __________ (1987), "Investment: Corporate Tax System and Capital Cost," in K. Hamada, M. Kuroda and A. Horiuchi ed., Macroeconomic Analysis of the Japanese Economy, University of Tokyo Press (in Japanese).

Takenaka, H., K. Takabayashi, Y. Horikoshi, Y. Kuwana and Y. Yoshida (1986), "The International Comparison of Capital Cost," Financial Review, No. 1, April, pp. 29-60 (in Japanese).

Teranishi, J. (1982), Japanese Economic Development and Finance, Iwanami, (in Japanese).

Turner, P. (1986), "Savings, Investment and Current Accounts: The Seven Major Countries' Experiences in 1965-84," Kinyu Kenkyu, Vol, 5, No. 3, pp. 1-48 (in Japanese).

Uzawa, H. (1969), "Time Preference and the Penrose Effect in a Two-class Model of Economic Growth," Journal of Political Economy, July/August (Part II), Vol. 77, No. 4, pp. 628-652.

Yoshino, N. (1984), "Composition of Financial Assets and Tax Free Saving Account," Gendai Keizai, No. 84, Autumn (in Japanese).

Zysman, J. (1983), Governments, Markets and Growth, Martin Robertson.

Investment, Finance and Corporate Tax Policy

G. J. Anderson[1]

Department of Economics, University of Toronto,
150 St. George Street, Toronto, Canada M5S 1A1

INTRODUCTION.

Throughout the postwar era there has been much concern in the western world about low productivity growth. The potential for embodying technical change and productivity improvememt in capital re-equipment has prompted economists to advocate policies that increase investment flows. These policies were intended to acheive this end via various tax incentives to agents that favourably influence their perceived cost of capital or cash flow position.

When the hoped for productivity improvements did not emerge governments were blamed, essentially because the structure of incentives, in discriminating between firms and agents, caused a misallocation of resources (see Daly, Jung, Mercier and Schweitzer (1985)). This would be characterised by tax wedges causing cash flow or capital cost variation according as diferent types of finance were employed for the same capital project. Similarly different types of factor inputs, in attracting differing tax burdens when performing essentially the same task, would be valued differently. Theoretically these constitute distortions in the way the market values real (as opposed to purely financial) activities causing a misallocation of resources if agents respond to market forces.

As an empirical matter this is an embarrasment for economists since these patently obvious theoretical arguements have not had such patently observable empirical counterparts! As Shapiro (1986) observes:

"One of the best-established facts in macroeconomics is that business fixed investment and output move strongly together over the business cycle. By contrast, investment and the cost of capital are either uncorrelated or only weakly correlated."

Bosworth (1985), Clark (1979) and Eisner(1978) are recent instances of researchers reporting difficulties in identifying significant tax effects in macroeconomic models of investment, though the issue has a much longer history (see for example the debate between Jorgenson (1971) and Eisner and Nadiri (1968)).

As Shapiro (1986) further points out these results are a problem for both neoclassical economists, because it apears that relative prices do not matter, and for neo-Keynesians, because it suggests that the interest rate does not influence the investment decision and an important part of the Keynesian transmission mechanism is lost. It is also a problem for both those who wish to advocate or criticise the implementation of discriminatory tax incentives since they do not appear to have much of an effect in either direction. Indeed the net effect of results thus far obtained suggests that governments need only consider revenue implications when evaluating corporate tax policy changes at the margin!

This paper discusses the implications of results from a research project which looks at the impact of the cost of capital and taxation in capital stock formation from a microeconomic perspective. The technical detail is reported in two discussion papers, Anderson (1987) and (1987a). Section 1 outlines the basis for and structure of the empirical equations and section 2 discusses the implications of the results, conclusions are drawn in section 3.

SECTION 1.

The cost of capital over a given period relates to the price paid for the use of capital resources being utilised, when multiplied by the cost of capital the result would correspond to the negative component in the firms cash flow that is associated with the maintenance of equipment to facilitate production. In the simplest ot all possible no tax, static "parable-like" worlds where q is the purchase price of a unit of homogeneous capital stock (K), r is the interest rate on a perpetual loan A over the appropriate period, ρ is the interest rate on a perpetual loan B over the same period (where A + B = qK) and δ is the rate at which the capital stock deteriorates over that period, the cost of capital, c is:

$$c = q(\gamma+(rA+\rho B)/(A+B))$$

If K is the desired constant size of the capital stock in this framework each period the firm must purchase γK for its maintenance and pay interest on the loan of $qK(rA+\rho B)/(A+B)$. Hence its negative cah flow is:

$$\gamma qK+rA+\rho B = cK$$

after its materials and labour costs (VC) have been deducted from its revenues (R) enough should be left to cover this amount if the firm is to be viable, hence

$$cK \leq R - VC \qquad 1.1$$

is required for viability in this state and firms choose a level of operations such that they can expect this condition to be satisfied. Add to this a perfectly competitive environment in the capital markets and the well known Modigliani-Miller separation between financial and real decisions will pertain (see for example Copeland and Weston(1983)).

Unfortunately the world is not this simple! The firms tax bill presents another negative component of cash flow. If this were a lump sum poll tax levied on the firm each period, viability requires:

$$cK \leq R - VC - T \qquad 1.2$$

If, somewhat more realistically, the tax bill is related to the firms net revenue (being for example a fixed proportion, θ, of it) and some proportionate relief, $\xi\delta K$, is given for replacement expenditures and μrA for interest payments on A type debt things become more complicated. By judicious choices of debt type and level of capital stock the firm can optimally influence its tax bill since:

$$T = (R - VC - \xi\delta K - \mu rA)\theta \qquad 1.3$$

Clearly one could write the cost of capital as a function of these various tax breaks so that it more apporopriately reflected the impact on cash flow of the capital stock decision. In this circumstance the Modigliani- Miller theorems no longer hold. More to the point the effect of the tax bill on cash flow or the cost of capital will vary to some degree with investment expenditures and one would expect that even in more realistic models varying the tax parameters would modify both investment and financial decisions. Yet, as has been noted, relationships between investment expenditures and taxation have been notoriously hard to establish empirically.

One reason for the limited success in observing relationships between tax policy and investment expenditures could be that virtually all the empirical work in this area has been pursued in a macroeconomic context. The complexity of corporate taxation and the variety of taxation arrangements that legally, industrially and geographically differing types of capital investing entity are subject to, make tax effects very firm specific. These could well be lost in the aggregation process concommitant with macroeconometric analysis. It follows that a microeconometric study may well be more appropriate in attempting to observe the impact, if any, of tax policy on investment expenditures.

Another reason for the failure to capture relative price effects in investment models is that simultaneity issues have been ignored in formulating cost of capital variables. Until now investment models have been formulated on the premis that a firms investment decision is separable from its financing decision in such a way that the cost of capital or the relevant cash flow concept may be considered independent ot the investment decision. Furthermore the statistical techniques used for examining such tax effects are predicated upon this assumption. However the tax bill the firm confronts depends on both the size of its capital stock and the level of debt interest payments it has to service, so that current investment and financing decisions jointly affect the impact of taxation on cash flow or the cost of capital. Hence the cost of capital is endogenously determined or influenced by the current investment decisions.

Infact the endogeneity or joint determination problem is more extensive than is immediately apparent. In the event of all investment expenditure tax allowances together with those for interest payments exhausting the tax capacity of the firm (the amount of tax it would otherwise pay), and in the absence of negative taxation, tax breaks to the firm on some aspects of financing and capital expenditures will cease to be operative and the cost of capital will increase substantially. As a consequence manipulation of corporate tax law which moves firms in and out of the "no tax payment" status should have a substantial effect on investment decisions making it even more puzzling as to why little or no tax effects have been observed empirically.

Whilst the ability to carry forward and backward these tax losses will lessen this effect in practice, it will still contribute to the mutual endogeneity of the cost of capital, finance and investment decisions. Furthermore a sybstantial number of firms in Canada regularly exhaust their taxbreaks, alternatively stated they do not pay taxes (for emample in 1982 over 45% of firms did not pay taxes), so that decisions contingent upon this switch between taxpaying and non taxpaying regimes could be important for a considerable number of firms.

THE INVESTMENT EQUATION

Anderson (1987) developes an empirical model of the capital stock in which the endogeneity issue related to the potential switch in the cost of capital associated with tax exhaustion is examined. Utilising a C.E.S. technology and assuming that financial structure is exogenously determined, an estimating equation for the ratio of capital stock to the firms total revenue w, for the ith firm may be derived as:

$$w = \gamma A(p/q)^{\alpha-1} \{(1-g)(d+r/(1-t))(r+\gamma)/(r+d)\}^{-\alpha} Q^{(\blacksquare-1)(\alpha-1)/\blacksquare} \qquad 1.5$$

where p is the price of Q, which is a unit of output; g is the rate of subventions and grants on new capital goods; r is redefined as the rate at which the firm discounts its future income stream; t is the corporate tax rate, d, the rate at which capital depreciation is allowed against taxes, α is the elasticity of factor substitution and $\blacksquare$ is the returns to scale factor.

The definition of r in this equation is complex, a simplified version which assumes that financial decisions are independent of investment decisions has been employed initially (but see below). Debt interest payments are tax deductible, hence when the firm (with a debt/equity ratio ℓ) is in a taxpaying position, the discount rate may be written as:

$$r_{Ti} = r_b (1-t)\ell_i /(1+\ell_i) = r_e /(1+\ell_i)$$

whereas in the non tax paying position it will be:

$$r_{NTi} = r_{Ti} + tr_b \ell_i /(1+\ell_i)$$

where r with subscripts b and e refer to the rates of return on debt and equity respectively.

By taking logarithms of 1.5 after some manipulation of the equation a structure for estimation emerges of the form:

$$\ln w_i =$$

$$C - \alpha[\ln(1-g_i)+\ln(d_i+r_{Ti}/(1-t_i))\}+(\gamma_i-d_i)/(r_{Ti}+d_i)] + \lambda \ln Q_i$$

if $TAX_i > 0$ and

$$C - \alpha[\ln(d_i+r_{NTi})+(\gamma_i-d_i)/(r_{NTi}+d_i)] + \lambda \ln Q_i \qquad 1.6$$

otherwise.

Note that allowable depreciation (d) is a measured variable in this context whereas economic depreciation parameters (γ) are unknown but can be estimated by industry. TAX is a variable indicating the tax paying status of firm i. Equations 1.6 are estimable if complimented by an equation describing the probability of firm i paying taxes, the model being one of the reggression switching regime variety with endogenous switching (see Maddala (1983))). Anderson (1987) developes a simple expectations formulation of the sources and uses of funds account which generates an equation characterising the probability of paying taxes.

Details of the results of estimating these equations and of the variables and data relating to tax information collected be Revenue Canada on over 16000 Canadian firms in the sample year

1982 are provided in Anderson (1987). Broadly summarised their implications are as follows.

With regard to the tax paying status of firms, those with low capital to labour ratios, where materials costs, salaries, wages and benefits constitute a significant part of the revenue disposition, tend to have a higher probability of paying taxes. This is not surprising since they are likely to be correlated with low debt equity ratios because, unlike investment expenditures, they do not provide the firm with items that can be used as collateral against debt. Profit retentions and book depreciation, when they form a larger portion of revenue disposition, are indicative of higher capital to labour ratios and debt to equity ratios. Since the disposition of revenues toward debt interest payments and investment expenditures are both tax deductable these variables are likely to be asociated with low probabilities of paying taxes. This is confirmed by the strong negative impact that these variables and the short and long term debt ratios have on tax paying status. Dummy variables indicating the type of corporation do not appear to have much influence on taxable status. However indicators of the type of industrial activity, reflecting disparities in tax policy across industries and indicators of location, reflecting disparities of tax policies across regions are all highly significant.

With regard to the investment equation itself a significant amount of endogenous switching is found to be present confirming that the cost of capital is not independent of the level of investment. In addition the cost of capital terms and their concommitant tax effects in the capital expenditure decision are found to be extremely significant, even in the presence of highly significant output effects. Furthermore dummy variables reflecting type of industrial activity and location do play a significant role in determining the probability of paying taxes, suggesting a considerable variety of tax wedges in the system.

FINANCIAL STRUCTURE

The forgoing has been pursued on the assumption that the financial structure of a firm was exogenous to the investment decision, that ℓ was an independent variable in the capital stock equation. Anderson (1987a) provides a preliminary investigation of whether the structure of finance is independent of technological and other firm specific considerations. Simple reduced form Tobit regressions of the shares of short and long term debt and equity in the overall financial structure as functions of technologically based and firm specific variables together with a profit retentions regression provide the basis for this work. The results suggest that in general technological and other firm specific factors are highly significant determinants of firm financial structure in a most plausible manner. Together with Anderson (1987) these constitute a substantial reversal of the usual implicit assumptions and results found in the macroeconomic studies cited above.

For various technical reasons, discussed in Anderson (1987), the magnitudes of estimates in the model may be overstated and much more work is required to verify the import of the results on other similar data sets. Furthermore the results on firm financial structure suggest that the interest rate variable is even more endogenous than has been allowed for in the investment model, However it is unlikely that the statistical significance of crucial parameters would be in anyway undermined by appropriate adjustments and with these qualifications they provide some extremely provocative insights to capital expenditure/ finance/ taxation issues in three major areas.

SECTION 2.

1. Tax rates and the investment tax credit.

It is possible to calculate firm specific response elasticities to various changes in tax law from the results in Anderson (1987). For example the elasticity of capital stock with respect to the corporate tax rate, for a taxpaying manufacturing firm in Ontario with an average debt - equity structure and facing a marginal tax rate of 45%, was approximately -2.2 so that a .45% increase in the tax rate would bring about a 2.2% reduction in its capital stock in the long run. Note that for a non taxpaying firm this effect would be 0, situation are more likely to be the larger capital intensive firms (see 2 below) so that in the aggregate the response may well be observed to be much smaller.

By calculating the "elasticity of the elasticity", the sensitivity of this elasticity to changes in variables that determine it can also be estimated. The appropriate calculations for variables such as tax allowable depreciation rates, the discount rate and the corporate tax rate are respectively .6436, -1.162 and -.0402. Thus while being reasonably sensitive to allowable depreciation rate and discount rate variations the elasticity alters very little in response to variations in the corporate tax rate.

On the other hand a corresponding calculation for the investment tax credit yeilded an elasticity of .52 so that scaling the investment tax credit by a factor of 1.01 would increase the capital stock in taxpaying firms by .52%.

These are representative of similar calculations that can be made for any variety of industry /province /firm tax status configurations generally with the same orders of magnitude. It is possible to conjecture as to why a reduction in the corporate tax rate has a larger impact than a corresponding increase in the investment tax credit for firms in a taxable position. In the former case long run corporate cash flow is affected on two fronts via investment expenditure write-offs (depreciation allowances) and debt interest payment write-offs whereas the latter case can be considered to correspond simply to an investment expenditure write-off.

2.Taxation and Scale Effects.

Returns to scale factors are significantly higher in non-taxpaying as opposed to taxpaying firms. Table I reproduced from Anderson (1987) highlights this result which is observed across all industry classifications and suggests that, given the eminently reasonable assumption that returns to scale are positively correlated with size, relatively smaller firms are carrying a disproportionate share of the corporate tax burden. It also suggests that tax policy has been rather successful in promoting investment where scale economies are apparent, though it says nothing about whether these are labour substituting or labour complimenting investment activities.

TABLE 1.

Industry	Returns to scale factor (constant returns = 1.0)	
	Tax paying firms	Non tax paying firms
Agriculture	.9999	1.0169
Mining	.9905	1.0002
Manufacturing	1.0058	1.0168
Construction	.9814	1.0000
Transport	.9956	1.0251
Wholesale	.9971	1.0196
Retail	1.0156	1.0262
Finance	.9837	.9890
Services	.9986	1.0252
Energy	1.0072	1.0260

3. Debt/Equity and Capital/Labour ratios.

An empirical regularity that emerged when analysing the data was that firms with high capital/labour ratios tend to have a low probability of paying taxes as do firms with high debt/equity ratios. This is not surprising given that capital investment and debt interest payments are both tax deductable. However, if the two ratios are positively correlated the so called tax wedge would be exaggerated. However the capital stock equation results discussed above are predicated on the simplifying and facilitating assumption that financial structure was independent of technology or that debt-equity and capital-labour ratios are unrelated.

Anderson (1987a) investigates this issue and finds the assumption lacks evidence in the data, a theoretical rationaliseation for this result is collateralising considerations. Since, in the absence of slavery, capital imputs rather than labour inputs provide a better form of collateral, debt (which is generaly the tax preferred source of finance) will be more accessible to firms with high capital-labour ratios and hence an independence assumption is unlikely to be valid. The theoretical arguement and its empirical support provide strong evidence for the notion that firms with high capital/labour ratios are more likely (and presumably find it cheaper and/or easier) to aquire debt than firms with low capital/labour ratios. There was also some evidence that subject to these collateral constraints, firms do tend to structure their finance in accord with the way in which it may be tax preferred.

With regard to internal versus external financing decisions (the profit retentions decision) monitoring or agency cost arguements (see Easterbrook (1984)) suggest a positive correlation between dividend payout and capital labour ratios and signalling arguements (see Bhattacharya (1979)) suggest a negative correlation. The data suggests that monitoring effects have more of an impact than signalling effects though such an observation requires so much qualification as to have little value at present. However sufficient information has been gathered to be able to establish a strong relationship between technological factors and financial structure.

Certainly this phenomenon is worthy of more attention and investigation. A more extensive study would develope a working model with more explicit links between the monitoring, signalling and taxation effects on corporate financial structure, tracking how these factors varied through time influencing and being influenced by technological considerations. Since the structure of financing is an important factor in cost of capital or flow of funds considerations, which in turn clearly impinges greatly on the investment decisions of firms, a clear understanding of these interactions is critical for any cogent debate on corporate tax policy. Note that this is a further rationale for the effectiveness of a corporate tax reduction relative to an investment tax credit increase noted in 1 above and exaggerates the discriminatory effects of such taxes in favour of capital intensive firms.

SECTION 3.

The import of these results is that cost of capital and tax policy implications for investment decisions are much more far reaching and significant than has been indicated by empirical research thus far. For example, it is clear that incentive structures implied by corporate tax policy currently encourage capital rather than labour intensive productive units. This may well be inconsistent with policy objectives asssociated with low unemployment targets since it would appear that a major tax wedge has been driven betwen capital and labour inputs, making the relative price of labour higher than is desireable. Furthermore these relative prices (or corresponding negative cash flow components) appear to significantly affect firms decisions. Consequently governments should concern themselves with much more than revenue considerations when evaluating corporate tax policy options.

Notes.

1) This work has been pursued as part of a contract with the Social and Economic Studies Division of Statistics Canada. Many thanks are due to Normand Lavoy for his dilligent research assistance. Helpful discussions with Brian Murphy and Michael Wolfson are much appreciated as are discussions with my colleagues in the research seminars at McMaster University, the Institute of Fiscal Studies in the U.K. and the editors and referees of the Journal of Public Economics and the Canadian Journal of Economics. Naturally all opinions and errors herein are the sole responsibility of the author.

REFERENCES.

Anderson G.J., (1987) "Is the cost of Capital Really Independent of the Level of Investment and Financial Structure of the Firm." Mimeo University of Toronto.

Anderson G.J., (1987a) "The Financing Structure of Canadian Firms an Empirical Note." Mimeo University of Toronto.

Bhattacharya, S. (1979) "Imperfect Information, Dividend Policy and the 'Bird in the Hand' fallacy" Bell Journal of Economics pp.259-270.

Bosworth B., (1985) "Taxes and the Investment recovery" Brookings Papers on Economic Activity 1:1985 pp.1-38.

Clark C.K., (1979) "Investment in the 1970's: Theory, Performance and Prediction" Brookings papers on Economic Activity 1:1979 pp.73-113

Copeland T.E. and J.F. Weston (1983) Financial Theory and Corporate Policy. Addison Wesley.

Daly M., J. Jung, P. Mercier and T. Schweitzer (1985) "A Comparison of Effective Marginal Tax Rates on Income from Capital in Canadian Manufacturing" The Canadian Tax Journal pp.1155-1192.

Easterbrook, F.H. (1984) "Two Agency-Cost Explanations of Dividends" American Economic Review v74 pp.225-238.

Eisner R. and M.I. Nadiri (1968) "Investment Behaviour and Neoclassical Theory" Review of Economics and Statistics v50 pp.369-382.

Jorgenson D.W., (1971) "Econometric Studies of Investment Behaviour: a Survey" Journal of Economic literature v9 pp.1111-1147.

Shapiro M.D., (1986) "Investment, Output and the Cost of Capital" Brookings Papers on Economic Activity 1:1986 pp.111-152.

The Optimal Reaction of Production and Investment on Uncertainty

Karl Aiginger

Austrian Institute of Economic Research, Box 91, A-1103 Vienna

1. AIM AND STRUCTURE OF THE PAPER

In this paper we show under which conditions uncertainty theory provides an unambiguous answer to the question whether firms in an uncertain environment will produce and invest more, the same, less than under certainty. In chapter 2 general models are developed which in principle apply to all decisions under uncertainty, though the decision variable is interpreted here mainly as production (for an overview see Aiginger 1987).

In chapter 3 we investigate parallel models about investment under uncertainty. In chapter 4 we use information out of the real world production process in industrialized countries to assess which models presented are more relevant.

The theoretical as well as the empirical results induce us to propose a new dichotomization of uncertainty situations into those of "petty" respectively "severe" uncertainty (chapter 5).

Though the results in this paper are arrived at by formal methods usually favoured by neo classical economists (expected utility maximization, density function as representation of uncertainty, micro level), the paper arrives at conclusions which can be interpreted as Keynesian in spirit (importance of flexibility, considerable impact of uncertainty on production, investment etc.).

2. A SURVEY ON MODELS OF DECISIONS UNDER UNCERTAINTY

In this chapter we will give a short overview on the models of decision of firms under uncertainty. Four general propositions are derived under which sufficient conditions are available to determine whether firms will produce more, the same or less than under certainty.

We will use Von Neumann-Morgenstern's Expected Utility Maximization. We will concentrate on passive models (where people can choose within a given framework without adapting it, information is given) and compare output decisions under uncertainty with output under certainty. The utility U depends on the variable Z (which could be understood as profits). Z itself depends on two variables X and Y (which usually are price and output). X is known under certainty (as X_0), in case of uncertainty a probability function about this variable - f(X) - is known. Y^+ is the optimal value of the decision variable resulting from the maximization in equation (1), $\bar{Y}$ is the optimal value of the decision variable in the corresponding uncertainty model (2).

$$\text{Max } U\,[Z(X_0, Y)] \rightarrow Y^+ \quad \text{(certainty maximum)} \tag{1}$$
$$\text{Max } E\,U[Z(X, Y)] \rightarrow \bar{Y} \quad \text{(uncertainty maximum)} \tag{2}$$

Proposition 1:

Linear technology ($Z_{XX} = 0$) plus $dY^+/dX > 0$ yields the following sufficient condition

$$U_{ZZ} \lesseqgtr 0 \rightarrow \bar{Y} \lesseqgtr Y^+ \tag{3}$$

Proposition 1 tells us that risk aversion may be a sufficient reason for a negative influence of uncertainty on the decision variable. This effect is often cited in literature, however the simple relation "risk aversion/neutrality/loving implies lower/equal/higher output" is correct only under two very restrictive assumptions. The first is that under certainty the optimum value of the decision variable, Y^+, depends positively on the value of X.

The second assumption is that profits are linear in the decision variable. This is the case in the competition model under price uncertainty, but not under a monopolistic model for an output setter with non-linear costs. In this case risk aversion may not suffice to guarantee a smaller output under uncertainty, risk neutrality does not guarantee that uncertainty does not change optimal decisions.

Proposition 2:

A linear utility function ($U_{ZZ} = 0$) and technological concavity, neutrality, convexity ($Z_{YXX} < 0$, $Z_{YXX} = 0$, $Z_{YXX} > 0$) yield the following sufficient condition

$$Z_{YXX} \lesseqgtr 0 \rightarrow \bar{Y} \lesseqgtr Y^+ \tag{4}$$

This proposition leaves aside risk aversion or loving, the effect of uncertainty now depends on technological conditions, like the cost and demand curve.

Up to now the models have assumed market clearing. Some variable adjusted ex post in a way to equal supply and demand. Equations (5 - 7) resp. (8 - 10) present a certainty model and a corresponding uncertainty model in which production y and demand x may differ, expected profits depend on the smaller of demand (x) or production (y) in equation (8).

Certainty model:

$$\pi = r(y) - c(y) \tag{5}$$
$$\pi_y = r'(y) - c'(y) \tag{6}$$
$$\pi_{yy} = r''(y) - c''(y) < 0 \tag{7}$$

Uncertainty model:

$$E\pi = \min\,[r(x), r(y)] - c(y) \tag{8}$$

$$\frac{\partial E\pi}{\partial y} = \underbrace{r'(y)}_{\text{marginal revenue under certainty}} - \underbrace{F(y)\,.\,r'(y)}_{\text{marginal costs of uncertainty}} - \underbrace{c'(y)}_{\text{marginal costs under certainty}} = 0 \tag{9}$$

$$\frac{\partial^2 E\pi}{\partial y^2} = r''(y)[1 - F(y)] - r'(y) - c''(y) < 0 \tag{10}$$

Proposition 3:

Given a certainty model of type 5 and an uncertainty disequilibrium model of type 8, uncertainty adds an additional marginal cost component which is positive (since $F(q)$ as well $r'(y)$ are positive). This yields for this type of model the unambiguous result of equation 11 (recall that $r''(y)$ is smaller than $c''(y)$ in the neighborhood of Y^+).

$$\bar{Y} < Y^+ \tag{11}$$

This proposition yields support for the above mentioned presumption of macroeconomists, that uncertainty will reduce output. Its most special case is where marginal revenue is constant: then output is maximized under demand uncertainty and a fixed price, a situation which could be labelled as "competition under demand uncertainty", as "uncertainty model with fixed prices" or as "stochastic rationing model" (Hymans, 1966; Malinvaud, 1980; Costrell, 1983; Benassy, 1983 and all the newsboy models in inventory literature). The unanimous result that production will be reduced stems from the expected costs of uncertainty: either production proves ex post to be higher than demand (implying high production cost the inventory cost may hopefully be reduced by further revenues from stocks) or production proves ex post to be lower than demand (implying foregone earnings and goodwill loss which may hopefully be reduced by the feasibility to backlog some part of unsatisfied demand). In any case expected costs are higher than under certainty. These extra costs of uncertainty - which are elaborated in Aiginger (1985) - are somewhat related to the arguments of "less-efficiency" and "noise signals" presented in the macro literature. Microeconomists however do not like models of this kind since model 8 assumes price stickiness in some ad hoc fashion and because the model is a partial model focusing only on the view of the producer under a given price (identical in the certainty and the uncertainty model).

A fourth channel for changing optimal production is given if it is possible to make a preliminary decision about the decision variable and then, after the veil of uncertainty is lifted, to revise this decision at some cost.

Proposition 4:

Suppose it is possible to make a preliminary decision $\bar{Y}$ and revise this upward (downward) at cost c_1 (c_2) then

$$c_1 \gtrless c_2 \quad \text{tends to imply} \quad \bar{Y} \gtrless Y^+ \qquad (12)$$

3. INVESTMENT MODELS

The number of feasible investment models is larger than of those for the output or price dicisions. What is added are the possibilities that the input of either one or both (all) production factor(s) must be determined before the veil of uncertainty is lifted, that this choice can be binding for output or not (fix or variable factor output relation), the production factors may be substitutable ex ante only, or ex ante and ex post (putty-clay assumption, putty-putty assumption), to which is added the impact of further attributes of the production functions (e.g. returns to scale). There may also be uncertainty - besides that on price and demand - in respect of the production factors' pay (wages) or their performance (and efficiency). Since long term considerations are especially important for investment decisions, the pressure to introduce dynamic models is greater than for output decisions.

Out of the theoretically feasible variety of models we present only a few, for a more complete overview and for the exact derivation of the results see Aiginger 1987 (chapter 7).

If both inputs are to be decided ex ante and prices are fixed, then the input decision is arrived at by the same laws as under certainty - though in respect to the output chosen under uncertainty. That means if the output under uncertainty equals the output under certainty, then the inputs are equal, too. If output is smaller (maybe due to proposition 1 or 2 or 3) then - given a well behaved production function - both inputs will decrease.

If capital has to be decided ex ante in a model of demand uncertainty (and disequilibria due to a fixed price), and labour can be chosen ex post (saving wages in case of low demand), then a linear version of proposition 2 applies. Whether more or less than expected demand is produced, depends on the relation between capital costs (which are lost in case of low demand) and unit profit (surplus of price over wages per unit).

If uncertainty about prices exist, but firms can sell at the market price whatever produced, we have an equilibrium model fitting to proposition 2. Let us assume that capital has to be chosen ex ante, labour ex post and there are no limits to the substitutability of capital, than we derive at the result, that investment depends on the relation between the elasticity of substitution and the degree of economies of scale. If substitutability is relatively low and/or if the production function approaches the linear homogeneous production function, capital input will be higher than under certainty:

$$\frac{\partial(F_{KL}/F_{LL})}{\partial L} \gtrless 0 \text{ if } \sigma \gtrless (1 - \mu)^{-1} \tag{13}$$

The economic interpretation of this outcome will be found in the fact that, when the factors can easily be substituted, a low capital input can be cheaply compensated for by means of ex post variable labour input; alternatively, given large diseconomies of scale, a contingently high price will lead to only a moderate increase of optimally profitable output. Generally it can be assumed according to the present model that capital input may be rather higher than under certainty. For example, given a substitution elasticity of 2, μ would have to be smaller than half in order to lead to a lower capital input. Not below a substitution elasticity of 4 might μ approach 0.75. For such seemingly realistic values of the scale parameter as 0.8 or 0.9, unrealistic substitution elasticities would have to be attained. If on the other hand the production function approaches linear homogeneous, then the possibility of reducing the capital input evaporates completely in this model of price uncertainty.

A minor change in the assumption can change the results. Kon assumes that the capital labor ratio has to be chosen ex ante. Ex post an optimal second step decision can be done, for less than full utilization of capital labour cost can be saved. The result now depend on a flexibility effect parallell to the last model, but a utilization risk effect is added which usually biases the investment downward. We have some sort of combination of proposition 2 and 3 and the results are ambigous.

Switching to dynamic models in principle does not change the story. If we model dynamic demand functions without disequilibria and without irreversibility we arrive at models of proposition 2 with - in general - ambiguous results. In Nickell's model for example optimal investment depends on the demand function with some more chances to bias capacity downward (though we know that other models would have given other results). If, however, we assume that investment can be revised upwards in case of strong demand, but downward changes of capital stock are limited to depreciation, we get a strong bias to a cautious dynamic investment strategy. The economic rational follows the spirit of proposition 3, namely that capacities may prove idle.

4. EMPIRICAL BEHAVIOUR

One of the critical facts theory points out and which has to be evaluated empirically is the degree of price stickiness in modern industrial production. Uncertainty will yield different decisions as compared with certainty if prices are sticky, albeit this has to be only a short-term stickiness. Only if prices are flexible in the sense of immediately offsetting demand shocks no disequilibrium costs have to be incorporated. We have demonstrated empirically by econometric and by survey methods that:

- industrial prices tend to reflect cost conditions and may be some price signals stemming from international markets, but do not react rapidly to demand shocks;
- prices seem to be less variable than quantities, especially the short-run fluctuations;
- price expectations are not less inaccurate as compared with production expectations (as they should be if the prices were an ex post control);
- price rigidities are at least as dominant in that sector of manufacturing for which a large number of enterprises, their small average size, etc., would suggest more competitive behaviour than for the rest; and
- asked about their response to a demand shock only 20 per cent of the firms in a survey cited price change as primary response.

On the other hand, quantities seem to be more flexible than assumed in most models, be it that production is an ex post control or be it that a preliminary production can be decided upon. Then, after demand is revealed, the decision variable can be partially adjusted (ex post flexibility); as indicators for a partial adjustment of quantities with a remaining part of disequilibria we found:

- the capacity utilization of industry as well as the inventory sales ratio fluctuates to a considerable extent, and surveys tell as that firms do consider these fluctuations as involuntary;
- output volume follows demand shocks closer than prices; and
- asked about reaction to demand shocks, 56 per cent of the firms labelled quantity changes a primary response, and 48 per cent reported changes in inventories (part of the disequilibrium is maintained).

In general, ex post flexibility seems to be greater than suggested in standard models. The distinction between ex ante variables, which have to be decided before the veil of uncertainty is lifted and ex post variables, which have to be decided (or which adjust) thereafter is not watertight.

However, ex post flexibility of quantities seems easier than that of prices (in contradiction to theoretical assumptions).

Ex post flexibility seems easier upwards than downwards. This seems especially true for input decisions, where this tendency for reported investment anticipations can easily be demonstrated on the micro and macro level. This asymmetry overcomes the theoretical tendency of uncertainty to increase optimal investment. From the theoretical viewpoint this source of bias is not very attractive. To model an asymmetry sounds ad hoc, if it is empirically true, the theoretical implications are trivial.

The mainstream model of competition with price uncertainty is exposed as an outsider in a real economy. Only 7 per cent of the firms consider it as relevant to their situation; a monopoly model with price as ex post control is chosen by 16 per cent. The remaining majority reported that prices are not an ex post control; quantity-setting and a partial ex post adjustment of the quantity set are considered as the most realistic models.

If we should draw a picture of a standard "representative firm" in modern industry we would do so in the following way: industrial firms have to decide on a preliminary production, they set up a cost price (or accept a market price). In case of demand shocks, output is partly adjusted, partly backlogs and/or inventories are changed, prices change slowly and in response to large shocks. Adjusting quantities upwards is easier/less costly than adjusting them downwards. At least sources three and four (marginal cost of uncertainty and asymmetric ex post flexibility) usually tend to bias (preliminary) decision under uncertainty downwards, technological concavity may add some other source of asymmetry.

Risk attitude is the most popular channel in the literature but most difficult to assess. If the results of entrepreneurs in a survey may be considered reliable, it looks as if entrepreneurs behaved as if they were risk-neutral for small, repeated decisions and risk-averse for large, one-shot decisions.

Using these findings (for detail see Aiginger 1987, chapter 6, 7, 8) for the production and investment model it looks as if models following the proposition 3 are of greatest relevance. For investment models in special we think that the technical possibilities to adjust the preliminary investment plan upwards and the necessary incorporation of the possibility that your capital stock may be underutilized will outweight the tendency shown in models of operationalization 2 that elements of substitutability will advise to provide a relatively larger capital stock under uncertainty to be able to meet eventually high demand.

5. "PETTY" vs "SEVERE" UNCERTAINTY

In general the result of the models invites us to find a new dichotomization of uncertainty situations. The dominant dichotomization of uncertainty literature into "risk" and "uncertainty proper" (according to the criterion whether probability functions about the uncertainty variable can be formed or not) is not very fruitful, since in the latter case only very crude rules of behaviour can be derived in a coherent and consistent way. The Keynesian view, that economic decisions are done in an environment much more complex than in an optimization problem where one certain variable is substituted by one for which a probability function is known, is nevertheless a useful warning. That no probability function can be assessed (or used implicitly) is an extreme alternative however, and precludes the economic analysis of a large area of economic problems.

We believe that it is important how the decision model is constructed, whether the importance of uncertainty will be considerable or minor, not whether we assume that probabilities can be assessed. If we construct models in which disequilibria exist and are not instantaneously closed by some ex post control, if we model the decision process as choosing between alternative techniques and degrees of flexibility, then we can use Neumann-Morgenstern's expected utility theory in general and probability functions and nevertheless describe a situation in which people behave "qualitatively differently" under certainty and uncertainty.

We tentatively propose that the real divide between uncertainty that matters and uncertainty with less consequences is whether there are chances to correct a decision (or at least to make errors in some way unimportant). This correction can either be a two-stage optimization process (short-run optimization for a given long-term optimization, e. g. for labour and capital), or it can be that the market price adjusts automatically yielding equilibrium for any quantity decision or that goods are durable so that unsold production can be used in the next period. We propose to label situations in which such adjustments are feasible as "petty" uncertainty, since the importance of uncertainty is mitigated to a large extent by these strategies. Models in which there are less strategies for ex post adjustments are labelled as "severe" uncertainty, since they usually result in disequilibria with important medium- or long-run consequences.

6. CONCLUSIONS

(1) We think that modern uncertainty theory can provide useful results about the effect of uncertainty on production and investment decisions. We derived four general propositions which can tell us about the impact of uncertainty depending on risk attitude, technological concavity, disequilibria and asymmetries in ex post adjustments.

(2) The effect of uncertainty depends on several characteristica of the models. If we assume equilibria in the sense that there is some variable (usually the price) which closes the gap between plans and the value of the uncertainty variable as revealed ex post, then the effect of uncertainty depends on circumstances not easy to evaluate empirically (third cross derivatives). If on

the other hand there are imbalances (production less or larger than demand, idle capacities) then the downward bias of production or investment under uncertainty is easy to establish.

(3) We follow during this paper the method of expected utility maximizing (EUM). This could be partly critiziced against the increasing amount of literature proving that individual behaviour does not follow EUM. We know about and accept this evidence, but we believe that most of our results could be even easier demonstrated if people do not follow this stringent logical system of EUM. Keynesians have always maintained that uncertainty just changes the rules, and that under "true" uncertainty (in contrast to risk) we should never use the assumption that people know probabilities. We claim that we can arrive at conclusions very near to Keynesian ideas by using the formal method of EUM. We can demonstrate the importance of flexibility, the fact that uncertainty significantly changes optimal decisions (even under risk neutrality). If we switch from the micro to the macro level it follows that growth and employment will depend under realistic conditions on uncertainty. It is the way in which we model the economy (equilibrium vs. disequilibrium, price or quantity adjustment, ex post parameters and their symmetry) which decides on the results and not the use of the formal method. If Keynesians accept this, they will be able to keep track in a world of mathematical economists (it remains a good stimulus if Post-Keynesians still maintain that uncertainty changes the model in an even more drastic way).

(4) To demonstrate the importance of the model chosen, we propose a dichotomization into "petty" uncertainty and "severe" uncertainty. Under "petty" uncertainty there is always a way to correct the decision after the veil of uncertainty is lifted and the economic consequences are not as large as under severe uncertainty. Severe uncertainty is characterized by important one shot decisions whith less possibilities to correct them afterwards. Economic consequences are large and there is a pressure to change the rules (the market, the behaviour, the production technique, the social system etc.).

Table 1

	Petty Uncertainty	Severe Uncertainty
	versus	
definitions	uncertainty is an intermediate problem, some variables have to be decided before the realization of X is known, some thereafter in a short run optimization or they adjust automatically	lack of ex post adjustments - no ex post control - price stickiness
characteristics	repeated (or small) decisions lack of serial correlation for realizations of X	one shot (large) decisions X serially correlated
	insurances	letal events (bankruptcy, dismissal)
	future markets	
	ex post flexibility, continuous adjustments	irreversibilities of investment and technologies
consequences	minor differences to certainty depending on facts difficult to evaluate (Z_{YXX})	important consequences (usually biasing down the optimal value of the decision variable); pressure to change the model to include new cost components and strategies
empirically testable conclusions for the relevance of the model	equilibria flexible prices and quantities uncertainty does not depress economic activity	disequilibria price stickiness uncertainty depresses economic activity

REFERENCES

Aiginger, K., "The Impact of Risk Attitude, Uncertainty and Disequilibria on Optimal Production and Inventory", Theory and Decision, 19, 1985, pp. 51 - 75.

Aiginger, K., "Production and Decision Theory under Uncertainty", Blackwell, Oxford, 1987.

Nickell, S. J. "The Investment Decision of Firms", Cambridge University Press, Digswell Place, 1978.

Symbols used

π	profit
U	utility
Z	argument of utility function (e. g. profit)
X	random variable; EX = X_0 (value of variable(s) under uncertainty)
Y	action variable
$\bar{Y}$, Y^*	optimal decision under uncertainty or certainty, respectively
U_{YXX}, Z_{YXX}	suffixes signify partial derivatives
p	price ($\tilde{p} = p + g$: extended opportunity costs = price + goodwill costs); $p \geq c$; $p \geq c'(q)$
c(q)	production costs (c' = marginal costs, c" = 2nd derivation with respect to q; c = constant unit costs)
x, q, s	demanded, produced, sold quantity
E	expectation operator
$f(\cdot)$	density function (symmetrical, smoothly differentiable);
$F(\cdot)$	distribution function of $f(\cdot)$, values between 0 and 1
r(x), r(q)	revenue function in dependence on demanded, on produced quantity
σ, μ	elasticity of substitution, scale paramter

Subject Index

MIX
Papier aus verantwortungsvollen Quellen
Paper from responsible sources
FSC® C105338

If you have any concerns about our products,
you can contact us on
ProductSafety@springernature.com

In case Publisher is established outside the EU,
the EU authorized representative is:
Springer Nature Customer Service Center GmbH
Europaplatz 3, 69115 Heidelberg, Germany

Printed by Libri Plureos GmbH
in Hamburg, Germany